Praise for Disaster Mental Health books by George W. Doherty

"Mr. Doherty has produced an invaluable reference volume for everyone involved in disaster response/disaster preparedness field. It is a must for your library! His attention to detail, breadth of scope, depth of knowledge and readable writing style, reflect the work of an eminent scholar in his field and one who has paid his dues on the frontlines. It represents the ultimate A to Z 'How to Do It' manual in this difficult, complicated field. From the sensitive discussion of clinical issues to the organizational planning details, the clarity and thoroughness of this volume are outstanding. This book should be required reading for everyone involved in this critical field."

John G. Jones, Ph.D. ABPP ATR-BC Licensed Psychologist

"Disasters happen—and someone has to be there to help the victims. George W. Doherty discusses training practices for mental health professional whose task it is to assist victims of disaster-related stress and trauma, giving advice and tips about dealing with various disasters whether they be the product of man or nature. His books are recommended to anyone whose career may take them into this type of profession and for any community library social science shelf."

—*Midwest Library Review*

"As a certified first responder with the City of Austin Emergency Measures Office I was delighted to find more information to add to my training. (The City of Austin provided training as a result of 9/11 in the event Austin, Texas, experiences a disaster from terrorists or other incidents of major concern.) George W. Doherty's book certainly presents a concise and informative addition to the library of a first responder, either beginner or one that is experienced. The information is well-researched and appropriate. Furthermore, I believe his books could be used by trainers when creating first responder to disaster training courses and be part of the study material."

—Irene Watson, Managing Editor, *Reader Views*

"This is an information-packed book about disasters and crises, the psychological impact of such events upon people, from the victims to the disaster workers, and also a psychological explanation of those who create crises, such as terrorists. Anyone who is considering being a disaster worker of any type, whether it is working for an organization like FEMA or even being an EMT, police officer, or volunteer fireman will find value in this book as it lays out various situations and what the disaster worker should know and be prepared to handle since an emergency or crisis could happen unexpectedly at any time."

—Tyler R. Tichelaar, PhD

"Awareness of how crises affect various people-groups, thinking through the important role disaster workers play in re-establishing normalcy in people's shaken lives, and planning immediate and long-term approaches to help traumatized people recapture mental equilibrium are vital aspects of a crisis intervention program. This is a beneficial and informative tool to raise awareness and plan levelheaded crisis intervention."

—Michael Philliber, PhD, for *Rebeccas Reads*

"It is extremely important for health practitioners and emergency workers to be prepared for emergencies, natural disasters, terroristic attacks and times of war. When dealing with traumatic incidents such as these, everyone is affected, including the victim, their family members and health care professionals. By being prepared to deal with these issues, research has shown that early intervention can reduce the chances of PTSD, acute anxiety, and depression. Having an operational emergency plan prepared ahead of time can make a huge difference in the ability to be prepared for the crisis.

"I found *Return to Equilibrium* to be very informative and necessary for mental health practitioners. As a person who works with disabled students in the community college setting, I also believe that this information is pertinent to college counselors and instructors. The better our understanding, the better we can serve their needs and help them reestablish equilibrium."

—Paige Lovitt, for *Reader Views*

Crisis in the American Heartland:

Disasters & Mental Health in Rural Environments

Coming Home: Challenges of Returning Veterans (Volume 2)

George W. Doherty, MS, LPC

Foreword by John G. Jones, Ph.D. and Alan Hensley, Ph.D.

eBook ISBN 978-1-61599-152-5
paperback ISBN 978-1-61599-153-2
hardcover ISBN 978-1-61599-154-9

Library of Congress Cataloging-in-Publication Data

Doherty, George W. (George William)
  Crisis in the American heartland : disasters & mental health in rural environments /
George W. Doherty.
      v. ; cm.
  Includes bibliographical references and index.
  Contents: v. 1. Introduction

  1. Crisis intervention (Mental health services)--United States. 2. Emergency
management--United States. 3. Psychiatric emergencies--United States. I. Title.
  [DNLM: 1. Crisis Intervention--United States. 2. Disaster Planning--United States.
3. Emergency Services, Psychiatric--United States. 4. Rural Health--United States. WM
401]
   RC480.6.D639 2011
   362.2'1--dc22
                              2011000004

The Rocky Mountain Region Disaster Mental Health Institute is a
501(c)3 Non-profit Organization. Learn more at www.RMRInstitute.org

Rocky Mountain DMH Institute Press is an Imprint of:
Loving Healing Press Inc.   www.LHPress.com
5145 Pontiac Trail   info@LHPress.com
Ann Arbor, MI 48105   Tollfree 888-761-6268
USA   Fax +1 734 663 6861

Dedication

I would like to dedicate this volume to two of my classmates who lost their lives fighting for our country in Viet Nam:

Thomas Holden, Lieutenant, United States Marine Corps
Classmate, St Mary's High School (Rutherford, NJ)
Graduate, United States Naval Academy

Glenn Bullock, Lieutenant, United States Air Force
Classmate, Pennsylvania State University Air Force ROTC
University Park, Pennsylvania

About the Cover (Front)

U.S. Army Chief Warrant Officer 2 Ryan Pummill, a UH-60 Black Hawk pilot with the 101st Airborne Division's 101st Combat Aviation Brigade, enjoys a light moment with his infant daughter after a homecoming ceremony in Hangar 3 at Fort Campbell, Ky., March 3, 2011. (U.S. Army photo by Sam Shore).
VIRIN: 110303-A-SG577-006

About the Cover (Back)

Family and friends hold American flags as U.S. Soldiers with the 219th Agribusiness Development Team, Indiana National Guard march through the doors of a hangar at Stout Field, Indiana, Oct. 20, 2010. The unit returned home after a yearlong deployment to Afghanistan. (U.S. Army photo by Spc. John Crosby/Released)
VIRIN: 101020-A-MG787-056

Army Lt. Col. T. Bradley Ninness, left, the commander of the 6th Battalion, 101st Aviation Regiment, 101st Combat Aviation Brigade, 101st Airborne Division, greets Soldiers returning from a yearlong deployment to Afghanistan at Fort Campbell, Ky., March 3, 2011. (U.S. Army photo by Sam Shore/Released)
VIRIN: 110303-A-SG577-001

Contents

Foreword by John G. Jones

As a retired military psychologist, I know I would have welcomed a volume such as this. A historical perspective is most important in understanding the development of improved treatment and intervention strategies for our returning veterans. Each of the multiple conflicts we have been engaged in impose new and different mental health challenges, and the current prolonged engagement is no exception. In fact, it is extraordinarily unique in many respects. Some of these are the multiple deployments, the heavy reliance on National Guard and Reserve units and the fact that a majority of the physical casualties have been from improvised explosive devices (IEDs). In addition, a great majority of the enemy contact has been in house to house combat and/or a result of enemy ambushes.

As in the past, each variation in the nature of the warfare exacts a different emotional toll on our troops. Further, in the many primarily rural states, the returning veterans will be isolated and far from readily available mental health resources. This poses multiple challenges for the mental health providers. As always, we must avail ourselves of the most current, effective and applicable treatment and intervention techniques. We also must work to insure that ALL our veterans in need of mental health care are indeed afforded it. These young men and women veterans have earned the right to have the best mental health care we as a nation can provide. I am convinced that this work by Mr. Doherty is a most valuable tool for the mental health professional to employ in providing this quality of care.

Mr. Doherty has a long history of dealing with trauma in a variety of settings and has published several other volumes in this genre. He is always extremely thorough, the information is well researched and apropos, and, perhaps most importantly, is readily applicable. I would strongly recommend that any mental health professional now working with our veterans or those preparing to do such work, include a careful study of this valuable work. It will stand you in good stead and prove to be an invaluable resource.

John G. Jones, Ph.D. ABPP ATR-BC
Captain USNR MSC Ret.

Foreword by Alan Hensley

Imagine yourself in a warm, comfortable, quiet place, devoid of distressing sights, sounds, smells, tastes, and feelings. You can reach out your arms or legs and touch the confines of your environment. Every aspect of your environment is known and non-threatening. You drift effortlessly without disruption. For nine months your environment is familiar. You become bonded and comforted by the voice of one person. Suddenly, at the end of nine months, you are inexplicably drawn into a constricting tunnel. The effortless floating you once knew now becomes scary and confining, unlike any event you have experienced in your short existence. Minutes or perhaps hours later, you emerge into a frightening world of bright lights, intense sounds, antiseptic smells, inexplicable tastes, and unrestrained feelings. Thus, begins the foundational experiences that will define your mental model, psychological contracts, and schematic architecture that will define how you view, understand, and respond to life experiences.

Passionate debate has ensued since the gleanings of attempts to understand human behavior. Evolutionists argued human psychology, personality, and behavior are an inherent artifact of evolution. Humans are born with inherent traits that predict our personality and behavior. Others, conversely, argue we are a product of education and experiences. We are born Tabula Rasa, devoid of any predispositional traits or tendencies. Conventional wisdom, however, holds that we emerge with predispositional tendencies, which are amplified or attenuated by learned conditioning and reinforcement. Through a process of education, experience, and role modeling, our mental model—the basis for viewing, understanding, and responding to life events—evolves from its rudimentary fragmented state to a structured, durable conceptualization of the workings of our world, as we understand it. Our mental model embraces our understanding of socially acceptable rules, roles, responsibilities, relationships, respect, morals, ethics, and values (R5MEV). Each life event we experience that coincides with these eight core attributes allows the individual's biological, psychological, sociological (biopsychosocial) and emotional state to remain in equilibrium. Events deviating substantially, however, result in disequilibrium, The comforting homeostasis of the person's biopsychosocial self, which facilitates the person to engage in long-term adaptive coping and planful problem-solving, is diminished in favor of dealing with actual or perceived crisis. The resulting emotional response diminishes rational and adaptive functioning of the person's prefrontal cortex, resulting in a largely innate short-term adaptive biological response, which is not necessarily long-term adaptive (or socially acceptable). To reconcile the emotional response, the individual must adjust the mental model to reconcile the offending event, completely restructure the mental model, or dismiss the event as illogical. Failure to do so results in continued, potentially lifelong distress—the basis of post-traumatic stress.

Concomitant with the evolution of the mental model is development of a myriad of psychological contracts. Every interpersonal relationship is navigated by a psychological contract, which is grounded in the same R5MEV that underpin our mental model. The first psychological contract every person constructs is the contract with our mother, which determines whether or not we emerge successfully from the first stage of Erikson's Model of Psychosocial Development. Successful emergence informs the trust we have for each successive person as we navigate life. Fortunately (or unfortunately), successful emergence, with positive reinforcement of successive early relationships predisposes us to inherently trust undeserving persons in our lives. Inherent trust predisposes us to the betrayal trauma, the most detrimental and insidious trauma etiology. While scars of physical trauma may heal, betrayal by ourselves or others destabilizes our mental model and invalidates psychological contracts and schemas developed over a lifetime that serve as a basis for viewing, understanding, and responding to life events. Deviation from accepted rules, roles, responsibilities, relationships, respect, morals, ethics, and values by self or others results in extreme biopsychosocial disequilibrium, resulting in an adverse emotional response.

The mental model and psychological contracts, which define our understanding of self in relationship to others, are enacted by a schematic architecture. Schematic architectures evolve from highly rudimentary at birth to highly durable and resilient through education, experience, and role modeling. As we observe others from birth, we learn how certain behaviors enable our survival. Evolving from effortful replication, such as using fingers to grasp, and feet to walk, to purposeful, rational adaptation, we develop schemas for seemingly effortless accomplishment emerging needs or desires. These actions that have evolved from purposeful cognition and development to automaticity enhance the probability of survival. Removal of a key attribute included in the schema, such as loss of an arm or leg, or even disfigurement, invalidates or diminishes the efficacy of the schema, resulting in biopsychosocial and emotional distress – hallmarks of posttraumatic stress. Specialization within the medical community results in a realm of "Ology." Instead of endeavoring to understand the etiology of PTSD at the holistic level, we in the psychological community attempt to assign highly constrained criteria, such as "The person has experienced, witnessed, or been confronted with an event or events that involve actual or threatened death or serious injury, or a threat to the physical integrity of oneself or others." In doing so, we fail to validate the factors that resulted in severe biopsychosocial and emotional responses, such as the self-inflicted death of Army Specialist Alyssa Petersen in Iraq in 2003. Threat of death or serious injury was not the causal element of her decision to remove herself from an environment of severe psychological distress. Rather it was compulsory violation of her established R5MEV upon which her mental model, psychological contracts, and schematic architecture was grounded. She felt helpless and hopeless to restructure, assimilate, or invalidate. The emotional response to severe biopsychosocial disequilibrium inhibited rational thought, resulting in her decision to expeditiously end the distress. The military not only has the highest suicide rate, the highest divorce rate, the greatest domestic violence involving military members and veterans on record. The causal elements are not what the individuals saw, heard, smelled, tasted, or felt. Rather, it is the inability to reconcile his or her behavior or the behavior of others.

In 2004, I performed a biopsychosocial evaluation of the Abu Ghraib Detention Facility abuses in Iraq. I found striking resemblance to the 1971 Stanford Prison Experiment conducted by Dr. Philip Zimbardo. A defense strategy I provided to the Area Defense Counsel analyzed behavior of the guards included the foregoing. The preeminent Dr. Zimbardo referenced the foregoing in *The Lucifer Effect: Understanding How Good People Turned Evil* (2007). Private Lynndie England, who was vilified for her participation in the atrocities, continues daily to experience severe PTSD as she attempts to reconcile her actions, as do the other defendants, while deployed in a gender role incongruent environment in the light of their "normal" non-military sociocultural environment. Unfortunately, England became the "poster child" for the atrocities, not because her behavior was more egregious than that of other defendants, but because she was a female engaging in non-stereotypical behavior, unacceptable for females in a non-deployed environment. Numerous veterans of Operation Iraqi Freedom, Operation Enduring Freedom, and even previous wars, struggle daily to reconcile the behavior of self and others in the light of the non-deployed environment. The result—over 22 suicides of active duty military and veterans per day.

In 2007, I presented "Why good people go bad: A case study of the Abu Ghraib Courts-Martials" at the Rocky Mountain Disaster Mental Health Conference, facilitated by George Doherty. Mr. Doherty has been an insightful, ardent proponent of understanding not only, that military veterans experience PTSD, but also why PTSD has become a hallmark of the wars in Iraq and Afghanistan. Unfortunately, current diagnostic and therapeutic techniques embrace the existent criteria in the Diagnostic and Statistical Manual of Mental Disorders that "the person [must have] experienced, witnessed, or been confronted with an event or events that involve actual or threatened death or serious injury, or a threat to the physical integrity of oneself or others." Jointly and individually, Mr. Doherty and I have observed the accepted therapeutic protocols that include such techniques as Eye Movement Desensitization and Reprocessing (EMDR) and Prolonged Exposure (PE) Therapy, which target the external manifestation of the cognitive distress, rather than addressing how the traumatic event violated the mental model, psychological contracts, and schematic architecture. Unfortunately, neither of these protocols assists the victim in understanding and compartmentalizing his or her behavior or the behavior of others in terms of the "dual hat syndrome."

As a practitioner and 26-year military veteran, I understand that proper screening, indoctrination and training precludes adverse biopsychosocial and emotional response. While persons entering the military are evaluated for occupational intelligence by such metrics as the Armed Forces Qualification Test (AFQT), emotional intelligence or personality domains and facets are not similarly assessed to ensure the individual's disposition and personality is amenable to the requisite duties. Also unfortunately, training and indoctrination of persons preparing for deployment is cursory and rudimentary, familiarizing military members only that they may experience adverse biopsychosocial and emotional response and identifying typical techniques for successful self-inflicted fatalities. In her final note, Specialist Peterson mused the suicide prevention training she had only days before received had provided her with the requisite knowledge to ensure that her final actions were indeed lethal rather than only debilitating.

A new patient appears in a physician's office with an abhorrent ulceration on his arm. Reacting to the manifestation of the evidence at hand, the physician asserts, "My, we need to apply and antibiotic and dress this." Within days the ulceration resolves, only to reappear after discontinuance of the antibiotic. The protocol is applied again, only to be followed by similar results. After several reoccurrences, the physician performs a hematological test, only to find the patient now has Stage III cancer that could perhaps have been treatable in the preceding months.

Medical and mental health practitioners are not innately adept by virtue of education or, in some cases experience, at understanding the etiology of military deployment-related PTSD. Rather than being a response to what the individual saw, heard, smelled, tasted, and touched, the etiology is reconciling these factors with the rules, roles, responsibilities, relationships, respect, morals, ethics, and values embraced by the individual. Not all apples are Red Delicious. Accordingly, therapeutic techniques cannot be addressed as "one size fits all." Rather, such variables as sex, gender, culture, and ethnicity must be foundational to the therapeutic technique. In his series, *Crisis in The American Heartland*, Mr. Doherty elucidates well on the need to engage in therapeutic techniques constructed upon these variables. Mr. Doherty also identifies that rural regions of the Midwest and other areas of the United States are ill prepared to address the emerging crisis posited to become multigenerational. We must endeavor to understand the foundations of PTSD rather than assigning contrite criteria to diagnose and eradicate externalized symptomology rather than addressing the insidious causal etiology.

Alan L. Hensley, PhD, Chief Executive Officer,
Prairie Oaks Trauma Care Center, Omaha, Nebraska

Preface

America has a history of pulling together and building a positive future which has been generally admired and looked up to by others throughout the world. When we have perceived a need to help out others, we have tried to live up to the ideals that have made this country great.

Many have referred to those Veterans who fought for this country in World War II as "The Greatest Generation". They are the ones that built the infrastructure and country that we have inherited and that we often take for granted. Too often we forget where we have come from, the sacrifices made by those who have gone before and the true legacies left by our predecessors.

I can remember as a young boy, standing with my grandfather watching as literally thousands of men marched by after returning home from fighting to victory in Europe and the South Pacific. We watched and cheered as Army, Navy, Marines, and even WACs, WAFs, Nurses and others passed by. We ran out of the house every time there was a formation of aircraft made up of bombers, fighters (then known as pursuit planes), cargo, and troop carriers flew over, destined for their home bases. I had uncles, family friends and others who told stories of their experiences while fighting or stationed in Europe and the South Pacific. Some of their stories I only learned about years later from others. The whole country welcomed them home. The whole country fully supported the mission they were on during the years of the war. For example, it took my grandmother a long time after the war ended to stop saving papers and scrap metal, etc. that was collected to support the war effort. There were many Sunday dinners when relatives and friends who had fought overseas gathered to talk and relax together. It was these Veterans who gave us the chance for the future we have today. They were selfless in their sacrifices and the country and the world benefitted from and thanked them for their efforts.

In the early 1950s, the "Forgotten War" in Korea involved many in our military. Once again, I was privileged to meet and know people who had served there. I had a more mature understanding of their efforts as most of them I knew and met as they were studying with me during my college years, thanks to a GI Bill that encouraged them to seek higher education and other trainings and supported them financially as they re-integrated into civilian life. Even though the official conclusion to the Korean War (officially a "Police Action") remains in limbo, most Americans welcomed and respected the sacrifices made by those Veterans during this conflict.

It was these men and women, relatives, friends, and others who influenced me strongly in my decision to seek a commission in the Air Force. I had wanted to do this since the days when I stood next to my grandfather as a boy watching our troops return home. During the Viet Nam Era, a large amount of respect for our military members and their families was somehow lost. Men and women who served during that period in any capacity were often mistreated by fellow citizens. In some cases, upon their return from combat or other duties, while in uniform, they returned to have things thrown at them, were spat on, discriminated against and generally mistreated. There was no real welcome home. It took years before any form of recognition was offered. I have friends who are named on the Wall in Washington, DC. One was a US Marine Officer who was a graduate of the U.S. Naval Academy and lost his life while leading his platoon in battle in Viet Nam. Another was a fighter pilot who was in my Commissioning Class through Air Force ROTC. There are many whose names are on the Wall which to this day are visited by friends, fellow Veterans and family to honor their sacrifice.

I remember starting my College years at a Commonwealth Campus. As such, when I transferred to the Main Campus I was told that ROTC (then still a requirement during the first two years) was only required for one year. In order to attain the Commission that I had wanted since childhood, I needed to press hard to get accepted into the Advance ROTC program. To accomplish this, I had to double up on my ROTC classes and extend my college years to an additional term, finally graduating and awarded a Commission in the US Air Force in March, 1964. There were many around me that disagreed with that decision. I did not.

The Draft (for military and other government service) was done away with following Viet Nam. Today we have a voluntary military. They are dedicated people who maintain a sense of service to our country. While generally supported by most people, many have little idea of what they do or why. Many have little contact with our military members. It is important that they be recognized and welcomed home and thanked for the missions they perform. They represent the legacy of those from "The Greatest Generation". Our Veterans who return home from such duties as Iraq or Afghanistan and other places around the world represent the hopes and future of the "Next Great Generation" for this country. They are the men and women who will build not just the infrastructure, but also the country that we wish to be able to pass on to those who follow us. We must not only welcome them home and thank them for their sacrifices, but we must also provide for and help them heal from the wounds they have incurred while upholding the values and ideals our country believes in. Our communities and families must work together to help integrate our civilian military members back into society where they can and will contribute to the future. As they come home, we all will find that a lot has changed. We must all find ways to return, not to a past "normal", but to an equilibrium that encompasses experiences, new learning, and a commitment to a better future for all. There is much to build and develop in our country. Our Veterans have the skills, motivation, will power, abilities and leadership skills to accomplish great things for our Next Generation. Let's not neglect the opportunity to accomplish this.

This book is the first of what will hopefully provide a background and some direction for those who work with our returning Veterans. They seek to provide ways to understand and promote understanding of their roles. Community leadership as well as informed counseling and active involvement are needed. For our wounded warriors, it is important to understand the wounds involved, physical and psychological, and to listen with open ears to what our Veterans are saying and not just try to apply a one-size-fits-all solution to what we might consider to be their situation. It is also important to include the families of Veterans in these approaches. While not perfect, I do hope that this book and others that follow contribute to a better understanding by all. This particular first volume is designed to increase awareness, review relevant history and research and open discussion about how best to apply previous lessons learned. As we move along, I hope that future volumes provide appropriate goals, methods and direction for all to achieve a new equilibrium. Comments, suggestions and contributions of ideas are always welcome. The future is ours to plan and to develop.

George W. Doherty
November 2012

Looking Back: A Snapshot from Rural America—1960

The alarm rang with a jingling sound. John rolled over and sat up in bed. It was 6 AM and he could see the light of dawn. The sun was not yet up, but he could see daylight over the hill to the east. It was time for morning chores. He heard his mother already down in the kitchen. He knew that he and she must do the milking and feed the other animals before breakfast. He d and went downstairs, saying good morning to his mother first. His father was gone for the week, having a job in the city and staying with his grandparents. Running the farm was up to him and his mother.

"Don't forget to put Pompeii in the milking station first, so she can be out in the pasture before the others," his mother called as he went out the door toward the barn.

John finished milking Pompeii just as his mother came into the milk house. He placed the milk bucket in the cooling area and moved out to feed the other animals and open the pasture gates so his 4-H sheep could move into their grazing area for the day. The sun was just beginning to peep over the hill as he and his dog "King" moved on to feed the chickens and ducks. There was a pleasant coolness with a touch of frost on the air. They went down to the pig pen where they were greeted with grunts as John emptied buckets of slop and other grain into the trough. As he headed toward the house, he and King played a little before putting some food in his bowl on the porch and some milk in a dish for the house cat Tippy.

Breakfast started with eggs and sausage. A bit of porridge and toast was just about ready as he saw his mother putting the milk containers into the milk refrigerator. She would transfer the milk later in the day into containers, after it had cooled. John talked with his mother while eating breakfast and she put together hamburger sandwiches and fruit for his lunch to take to school. "Got all your homework from last night and your books for today?" she asked.

"Yes," John replied. "I also have my library books to return and my fishing pole, so Phil and I can fish in the creek during lunch and recess."

"Good," his mother said. "You best hurry. You don't want to miss the bus." John finished his breakfast, went upstairs to change into school clothes, gathered up his books and lunch box and fishing pole, kissed his mother goodbye, and told her he would be home after 4 PM as it was Monday.

John and King ran and played a bit heading down to the dirt road along the creek that led to the main road. John told King to go home when the road and creek curved around to enter the woods. King yelped and obediently went home as he always did. John enjoyed his one-mile walk out to the highway. The sun was now up and the frostiness was lifting. The trees were just beginning to show their fall colors. A deer ran off to his left. He reached the road intersection with time to spare. In about ten minutes, the little blue school bus showed up. The other students onboard all said "good morning" as John got on board. He returned the greeting. There were only about seven or eight students on the ten-mile trip to their school house. The school itself only had 25. There were two classes in each room and they shared books, blackboards, and alternated activities. When one grade was doing activities and written work, the other was getting instruction. All the children were from farm families and had similar chores and work as did John. During recess and lunchtime, they fished in the nearby creek or played baseball. All belonged to the County 4-H Club and often helped each other out on their projects. During hunting season, they all took a week off for hunting and also took a week for the harvest each fall.

Community get-togethers took place during harvest time, Thanksgiving, and Christmastime. Fourth of July was celebrated with a picnic with all bringing food for everyone, dance, music, and games. At other times, they visited each other's farms and even had a small astronomy club. Few farms had phones, some had early televisions. All had radios and got messages through radio programs and enjoyed the early Saturday morning radio polka shows with requests and dedications for each other. John graduated in a class of eight students. He went on to high school, got a degree in Agriculture from the state Land Grant University and served as an Officer in the U.S. Air Force. During that period, much changed in the world. However, rural communities and frontier areas have maintained many of their same values, but also lag behind in some of the benefits of this modern, changing society—some for the good and some with new challenges and crises.

PART I:
Background and Overview

Chapter 1: Why Community Leadership Is Important

"We can't all be heroes, because someone has to sit on the curb and clap as they go by."

Ralph Waldo Emerson

Throughout history, men and women have changed the world through their strength of will and courage of conviction. Such extraordinary people find something within themselves to change the way things are. They are able to inspire others to follow them, despite the odds. In the rotunda of the Jefferson Memorial in Washington, D.C. are inscribed the following words penned by Thomas Jefferson: "We hold these truths to be self-evident, that all men are created equal... endowed by their creator with certain unalienable rights."

While we can marvel at the vision of a Walt Disney or a Henry Ford, who created new ways of entertainment and transportation, there have been only a handful of people in history who have had the audacity to declare a new state of human rights. Prior to the words written in the American Declaration, people did not have rights. Kings had rights. There was no special privilege to just being human. It was an extraordinary act of courage to declare such a new reality. It inspired men and women to action and ultimately to the action of war. People voluntarily gave their lives because they yearned for freedom of expression, found it, and declined forever to forfeit it.

The feeling of passion evoked by the Jefferson Memorial represents a standard of leadership. True leaders, through the integrity of their values, communication, and action, have been able to inspire others to act differently. They accomplish this, not by forcing them to act differently, but by creating a clear glass for them to see themselves, their deepest desires and finest motivations, and act accordingly.

Goals

Every organization and team is organized to provide certain services and has responsibilities outlined in their charters, protocols, or statements of purpose. Team goals are important. When team members work together and share ideas and responsibilities, they can accomplish much more than a group of individuals each working alone. Keeping everyone working toward the team goals may require close attention and occasional redirection in order to maintain the team's purpose and focus, and to accomplish their mission. It may be necessary to occasionally remind members of the team's purpose. It is easy to get sidetracked or lose sight of the goals. A good leader provides encouragement and motivation by showing appreciation for good ideas and extra effort. Mediating differences and disagreements between team members by stressing compromise and cooperation helps keep people on track and goal-oriented. Involving team members in discussions and decisions allows everyone the opportunity for input. However, there are times, as the team leader, when you must make the decisions by yourself.

Getting to Know People

People have different abilities, needs, desires or wants, and purposes in life. In order to get results and get along with others, you must find out what makes them tick. There are a number of ways to accomplish this. They all involve meeting and getting to know others in order to work effectively.

1. Interact with other team members as often as possible. Usually the best way to get to know others is through direct personal contact. Take note of each person's unique qualities and characteristics.
2. Treat others as individuals. Use your knowledge and understanding of each team member.
3. Be aware of expectations. Each team member deserves individual treatment. Everyone expects something different. Some expect recognition. Others expect an opportunity to learn, a chance to work with others, etc.
4. Provide reinforcement or rewards for performance. A pat on the back is always a source of personal satisfaction and is a positive reinforcement for a job well done.
5. Delegate responsibilities. Team members should share in the work to be done. That way, everyone can also share the pride in the

accomplishment of the team. In this regard, it is also very important that each team member know what is expected of him/her, what resources are available, what deadlines need to be met, etc. It is important that delegated team members accept responsibility for getting things done. This is an aspect of leadership in which everyone can and should excel.

As a leader, you should become actively involved. You are not able to do the job all alone. However, you can help get the job done better and faster by:

1. Taking the initiative. Do not stand around and wait for someone else to get things started. Roll up your sleeves and dig in.
2. Seeking help and information. If you need advice, don't hesitate to ask for it. This encourages group involvement and helps accomplish group goals.
3. Offering information and help. It may be that your particular skills and knowledge are exactly what is needed.
4. Knowing when and how to say NO. When your time and/or resources are already committed, it is OK to turn down extra tasks. However, it is important to do this tactfully and politely.
5. Making things happen. Be decisive, energetic, and enthusiastic. By doing so, you increase the probability of getting things accomplished.

Leadership Values

Leaders are responsible for making sure they have the right rules (i.e., those policies that reflect their values) and that they carry them out themselves. If responding is a value, then they can't be sitting behind their desks in times of crisis. They need to be active, on their phones, in the community or involved together with others who also devote their lives to this value. It will be critical that our community leaders and other prominent citizens be actively involved as leaders, for example, in working with our Veterans as they return home. Our returning Veterans also will be prime candidates as leaders in our communities and in business and industry. These men and women already have training in various skills and abilities. They are also well trained in leadership skills through training and experience. As such, they already have much of what we need in these ways in our communities and other areas. They have the opportunity to rival many of the accomplishments of "The Greatest Generation" that returned from World War II and built our country into a well-respected one in industry, business and world power in a very positive way.

We all want to believe in an ideal. While no leader is flawless, the good ones take on the task of trying to live up to the ideal. In our age of speed and change when nothing seems stable, a leader's role is even more important. After more than two decades of writing about the qualities of leadership, Bennis (1999) believes it comes down to "... in tomorrow's world, exemplary leaders will be distinguished by their mastery of the softer side—people skills, taste, judgment, and, above all, character." (p.19).

Everyone wants to work with people of impeccable character. Just as one's own character determines one's personal ability to generate trust, so it is for the organization or community as a whole. An entire organization built on the strength of character as its foundation would be a compelling place to work. Such an organization requires focus on three aspects: responsibility, integrity, and generosity of spirit.

Responsibility

Many organizations, like many individuals, have public values. Being responsible for them means interpreting them as spirit rather than law. That way, the values become a positive guide rather than a hindrance to be avoided. To see values in this way is seeing them not as policy, but rather as a body of ethics that created policy in the first place. Politicians are often heard saying things such as "We play by the rules, and we all know that we need to change the rules." An admission such as this suggests that the body of ethics that guides behavior, the character of the individual, is subservient to the rules. It further suggests that we interpret the rules to avoid our own responsibility rather than applying them as positive guides for our action. Such an approach is unconscionable. It will inevitably ring hollow and will not inspire any measure of loyalty or commitment from your people. Responsibility doesn't mean never changing. However, if change is proposed, it should be consistent with one's fundamental beliefs. One should be "able to respond" to what is needed without altering the basis of who one is as a person and organization.

Integrity

The second aspect of character is integrity—actually doing what one says one will do. It is the follow-through to being responsible.

Generosity of Spirit

The final aspect of character is generosity of spirit, which has two dimensions. The first is being able to maintain a perspective beyond what obviously serves one as an individual—or even beyond what obviously serves one's organization. The second is a graciousness of attitude about one another. Developing a broad perspective results in caring for all stakeholders, including the greater community.

The second dimension of spirit, graciousness of attitude, is critical because it manifests itself daily, minute by minute. Situations may seem ambiguous in a world moving as fast as ours. It is important to provide the context of change as a way of mitigating the ambiguity. However, providing context is not always possible. Many times, we need to accept each other on faith. Graciousness may sound soft. However, it speaks to the heart of teamwork and support for one another, even as one discusses, disagrees and makes tough decisions. One may want to foster the ability to presume the best about fellow workers, clients, and others. Graciousness means that one willingly places trust in others, generates good will, belief, and faith in what is good, rather than focusing on failure or threats to one's own existence.

Conclusions

With every decision one makes, one risks an unknowable result. Regardless of how fine the analysis, the future will stubbornly remain unknown. What is not obvious is that the chasm between what one knows and what one can't know has to be bridged by an act of faith, and any act of faith will attract naysayers who demand more evidence. Such cynicism erodes relationships and freezes action. When character is developed, decisions are based on the assumption that people will respond positively when they are trusted, supported, and cared for.

In our current society of electronic everything, it would be easy to think that leadership is about information sharing or moving more quickly than some other person in one's field. However, precisely because the world is electronic, a leader's job has shifted to one of personal, values-based communication and action. It is easy to get caught up in the tactical day-to-day operation of the organization. It is easy to move from graciousness to selfishness. Responsibility demands that one spends a lot of time considering what one stands for, acting in a way consistent with the answer, and then making sure that the principles of behavior are reflected in one's organization.

It is difficult for leaders to be so visible and be held to a higher standard than the rest of the workforce. However, it is also a privilege and an incredible opportunity. Knowing what to do and constantly looking for ways to improve is challenging, rewarding, and provides an atmosphere where everyone wants to contribute—and usually does. Our returning veterans have been trained in these values, exposed to adversity, and are prime candidates for leadership positions in our communities. Together with the current leaders in our communities, our veterans can and will live up to their legacies and have the potential for being the Next Great Generation.

References

Bennis, W. (1999). *The Leadership Advantage. Leader to Leader*. No. 12, Spring.

Collins, J.C. & Porras, J.I. (1994). *Built to Last: Successful habits of visionary companies*. New York: HarperBusiness.

Locke, J.L. (1998). *The de-voicing of society: Why we don't talk to each other anymore*. New York: Simon & Shuster.

Morrow, Shaw, R.B. (1997). *Trust in the Balance: Successful Organizations on Results, Integrity, and Concern*. San Francisco: Jossey-Bass.

Naisbitt, J. & Aburdene, P. (1990). *Megatrends 2000: Ten new directions for the 1990's*. New York: Morrow.

PART II:
Morale, Deployment and Stress

Chapter 2: Brief Background and Review

"I have never accepted what many people have kindly said—namely that I inspired the nation. Their will was resolute and remorseless, and as it proved, unconquerable. It fell to me to express it."

Winston Churchill, on his 80th birthday,
address to Parliament 11/30/54

What is Morale?

Manning (1991) defines morale as a function of cohesion and *esprit de corps*. There is a distinct national character for each nation. It affects the ways in which an army fights and can affect the outcomes of battles. Knowledge of the enemy's national character can help commanders and planners determine the outcome of any given battle (Labuc, Stasiu, 1991). Stewart (1991) examined the cohesion, morale, motivation, and unit performance of both sides in the Falklands conflict. Based on face-to-face interviews with British and Argentine officers, NCOs (noncommissioned officers), and enlisted personnel, she explains the successes and failures of land forces during the 1982 campaign. She shows clearly that cohesion is indeed a "force multiplier" and in many instances determines small units' ability to stand and fight.

History

Seidule (1997) examined the morale of the American Expeditionary Forces during World War I to determine its competence. When examining morale, military psychologists cite the importance of three factors: cohesion, esprit de corps, and biological and psychological needs such as health, rest, and nutrition. Morale, however, has more to it than the battlefield determinants, particularly in World War I. Societal values also played a key role. Officers and soldiers came into the army with idealistic and romantic expectations of war and service. Those expectations were particularly strong because the army used conscription without sufficient training to overcome pre-war perceptions. The morale of the AEF was poor. Poor morale served as both an indicator of inferior battlefield performance and a factor in the Americans' tactical problems. The sorry state of the AEF in November, 1918 would probably have precluded it from continuing to fight into Germany in 1919. Studying morale provides an evaluation of an army at the tactical level by assessing more than just tactics. In the case of the AEF, poor morale was a cause and a symptom of tactical ineffectiveness (Seidule, 1997).

Keene (1994) reviewed the broad agenda pioneering military psychologists during World War I set for themselves in studying problems relating to the psychology of the soldier. Psychologists in the newly created Morale Division, for example, initiated rudimentary studies on issues that would become staples of 20th-century military psychology. These included the adjustment of recruits to Army life, the effectiveness of Army propaganda in changing attitudes, the reasons soldiers' desert, and the impact of military service on civilian soldiers. By encouraging military policymakers to consider the importance of soldier psychology in a mass army, the Morale Division introduced a new and somewhat controversial perspective into the organization. Consequently, plans to develop soldier morale were integral to the interwar mobilization plans prepared by the Army War College, considerations notably absent in 1917.

From the outset of World War II, Canadians wanted to avoid the horrors encountered on the western front in 1914-18. The most significant was "shell shock". Most medical personnel preferred not to assign to combat those who showed neurotic symptoms during training, but this approach was challenged by the Canadian Psychological Association and by the new Personnel Selection Directorate established in 1941 (Copp & McAndrew, 1990). Personnel Selection claimed to be able to distinguish, before training, between those suited and those unsuited to combat duty. When Canadian troops went into battle in Italy, however, the preparatory work seemed to have had little impact. Canadian losses due to "battle exhaustion" were no less than those of other allied forces. Front-line treatment allowed about half of those affected to return to their units. Eventually, however, a very large number of soldiers were assigned to non-combat roles because it was judged they could no longer function effectively in battle. Copp and McAndrew (1990) are critical of military commanders who thought strict discipline coupled with high morale from good training and success in battle would keep battle exhaustion in check, and of officers in the Royal Canadian Army Medical Corps who tried to impose theoretical solutions that did not fit the circumstances. The authors suggest that some doctors, using energy and common sense, contributed to the evolution of contemporary psychiatric ideas about the realities of large-scale psychological casualties.

Suicide can sometimes be the result of lowered morale, but also may be a result of training and expectations. Suzuki (1991) discusses two facets of an American program to prevent suicide among the Japanese during WWII. One was a research component in the foreign Morale Analysis Division (FMAD), a subunit of the Office of War Information. The principal FMAD researcher on Japanese suicide and ways to prevent suicide among the Japanese military was the anthropologist R. Benedict, assisted by a Japanese-American aide, R. Hashima. The second facet was the suicide prevention program itself, put into effect toward the end of the war in the battles of Saipan and Okinawa. American GIs used Allied propaganda in an attempt to alter the professed no-surrender policy of Japanese military leaders.

More Recent Changes

U.S. military forces are increasingly involved in a variety of multinational peacekeeping and humanitarian assistance missions. In preparing soldiers for these, it is critical that leaders have an understanding of the nature of the stressors involved. Since the end of the Cold War, nontraditional military missions have increased substantially, whereas armies are being downsized. Survivors of downsizing are being called to do more with less. One example was the unscheduled 1994 deployment of a Patriot missile battalion to Korea (Segal, Rohall, Jones & Manos, 1999).

Between 1989 and 1996, the active duty Army was cut from roughly 770,000 to 500,000 personnel with more cuts to come, McCormick (1997) evaluated the manner in which the Army has adapted to and been affected by this externally-mandated organizational change. In examining the process of downsizing, he considered several dimensions of military effectiveness. He examined the Army's political effectiveness—the ability of its leadership to articulate the Army's role and to obtain needed resources in an increasingly treacherous political and budgetary environment. He also evaluated the Army's organizational effectiveness, focusing particularly on how the Army personnel community managed the downsizing of the officer corps. He considered the Army's objectives in this process, the appropriateness of these objectives, and the Army's success in achieving them. Finally, he evaluated how downsizing has affected the morale, commitment, attitudes and behavior of the Army officer corps, intangible yet crucial aspects of military effectiveness. The downsizing of the Army is a story of both failure and success. The Army's leadership failed to make a persuasive case to civilian leaders for a larger force and the resources necessary to maintain it. Consequently, other aspects of military effectiveness have been jeopardized. Conversely, the Army's leaders successfully planned and implemented dramatic personnel reductions, particularly within the officer corps. The Army achieved its downsizing objectives and these objectives were for the most part appropriate. But, despite the Army's best efforts, prolonged and incremental downsizing has taken its toll on the officer corps, undermining morale, commitment and professionalism, and perhaps with this the Army's ability to fight and win future wars. The Army's outdated officer management system and the legislation that governs it have exacerbated these undesirable effects. McCormick (1997) concludes with suggestions for the reform of these systems in light of the unprecedented challenges brought about by the post Cold War era.

Contemporary organizational theory consistently argues that the structure of an organization can affect the behavior of its members. Research suggests that members of organizations structured in a rigid hierarchy exhibit lower morale, decreased innovation. and lower output when compared to organizations structured less rigidly (White, 1998). While the research is clear on the impact of hierarchical design on overall behavior, it is less clear on the effect that rigid hierarchy may have on ethical behavior. White investigated the hypothesis that rigid organizational hierarchy inhibits ethical behavior. Ethical behavior was operationalized by applying Kohlberg's (1976) model of six moral development stages, and measured by Rest's Defining Issues Test (DIT).The hypothesis was tested by comparing the mean DIT scores of 480 Coast Guard personnel with the means of meta-samples of individuals from less rigid organizations. A military organization, the U.S. Coast Guard is a stereotypical rigid hierarchy, with a tall pyramid structure, numerous hierarchical levels, centralized decision making, and an emphasis on obedience and heteronomous behavior. The hypothesis was tested further by comparing DIT scores of Coastguardsmen assigned to large ships with Coastguardsmen assigned to shore units. Because of the extremely regimented routine aboard ships, sailors are allowed little opportunity to act autonomously. It was hypothesized that the DIT scores of seagoing Coastguardsmen would be lower than their counterparts assigned to less rigidly structured shore units. The findings supported the hypothesis that a rigid hierarchy restricts morale development. Coast Guard respondents scored about seven points lower on the DIT than a large adult meta-sample from society-at-large. Sea-based Coastguardsmen scored significantly lower than their shore-based counterparts. This suggests that the extremely rigid hierarchy of the shipboard environment further restricts morale development. This study suggests that an ethical dimension be added to the negative consequences of rigid hierarchy. It also argues that the military, from a morale development perspective, is different from mainstream society, and that the difference is caused primarily by the rigid hierarchy employed by the military. The findings reinforce the need for democratic, civilian supervision of military policy and provide a recommended strategy for accommodating the tensions between rigid hierarchy and morale development.

Deployment Factors that Affect Morale

The U.S. military is increasingly involved in operations that require specially configured task forces that are tailored to the demands of a particular operation. Given the presumed importance of unit cohesion as a social influence on morale, performance, and stress resiliency, a critical question is how cohesion develops in such units. Bartone & Adler (1999) examined cohesion over time in a U.S. Army medical task force that was newly constituted to serve in a United Nations peacekeeping operation in the former Yugoslavia. Survey data from 3 phases of the operation (predeployment, mid-deployment, and late-deployment) suggest that cohesion levels develop in an inverted-U pattern—starting out low, reaching a high point around mid-deployment, and then decreasing again toward the end of the six month mission. ANOVAs comparing work groups or sections within the task force revealed group differences on cohesion, with military police and physicians highest and operating room staff (nurses and technicians) lowest. Situational and home environment stressors correlated negatively with cohesion during predeployment, whereas work relationship problems were stronger (negative) correlates at mid- and late-deployment. Results demonstrate the importance of assessing cohesion across phases of deployment.

Segal et al (1999) present data drawn from surveys of soldiers in the battalion that replaced them. Although both PATRIOT battalions were quite similar in many respects, the soldiers in the first battalion sent to Korea had been told that they would not be deployed again for 2 years, had less warning of their deployment, and had seen more deployments than the second battalion sent to Korea. In both battalions, the best predictor of morale for younger soldiers (E4 and below) was family adjustment to Army life. The best predictor of morale for older soldiers (E5 and above) was leadership support for soldiers. Data revealed that both junior and senior enlisted soldiers in the first battalion had significantly lower morale and family adjustment ratings than the soldiers sent to replace them. Findings reinforce the importance of communication with the survivors of organizational downsizing and consideration of the needs of their families as their jobs undergo restructuring.

Bartone, Adler & Vaitkus (1998) summarize findings from a longitudinal case study of 188 U.S. Army soldiers (the group was 78% male) in a medical unit performing a peacekeeping mission in the former Yugoslavia. The goal was to identify key sources of stress and to delineate the effect of these stressors on the health, morale, and mental readiness of soldiers. Findings suggest a range of psychological stressors varying across operational phases of a peacekeeping mission. The degree of stress experienced in various areas correlates significantly with depression, psychiatric symptoms and low morale. The range of stressors is summarized in a model of five underlying dimensions of psychological stress salient to soldier adaptation in peacekeeping operations: isolation, ambiguity, powerlessness, boredom, and danger/threat. This model suggests several recommendations for countermeasures that organizational leaders can take to maintain soldier psychological readiness during peacekeeping operations.

"Commuter Wars"

Richard & Huffman (2002) note that the US Air Force has developed a military force that can fight by night and return home by day. This phenomenon of "commuter war" was especially evident during Operation Allied Force over Kosovo. 540 military personnel participating in

Operation Allied Force were administered a survey measuring morale, wellness behaviors, and work-family conflict. The deployment had adverse effects on wellness behaviors of permanent party and temporary duty assignment populations. Additionally, levels of morale and motivation varied between the two groups. Permanent party personnel also reported increased rates of work-family conflict. Results suggest that commuter war affects wellness behaviors, morale, and work-family conflicts of military personnel.

Environmental Factors

Bliese & Britt (2001) examined the degree to which individuals' reactions to stressors were influenced by the quality of their shared social environments. Based on social support theory, they proposed that individuals in positive social environments would show lower levels of strain when exposed to stressors than would individuals in negative social environments. The quality of the shared social environment was assessed by measuring the degree of consensus among group members about an issue of importance to the group—namely about the group leadership. Social influence theory provides compelling reasons to believe that this measure of consensus should be a strong indicator of the quality of the social environment within the groups. In multilevel analyses using a sample of 1,923 soldiers who were members of 52 Companies deployed to Haiti, Bliese & Britt found that the quality of the social environment moderated relationships between (1) work stressors and morale and (2) work stressors and depression.

Stress Factors

High levels of stress have been associated with morale and well-being issues among soldiers. Micro factors such as coping and macro factors such as leadership environment and group cohesion have influenced stress levels of soldiers in the workplace. Soldiers' self-report of stress levels in relationship to soldiers' perception of leadership, group cohesion and coping was investigated (Arincorayan, 2000). The amount of stress experienced by a soldier was measured by the Brief Symptom Inventory (Derogatis, 1977). Soldiers' perceptions of leadership environment were measured by the vertical cohesion scale (Walter Reed Army Institute of Research, 1996). Soldiers' perceptions of coping method were by the emotion-focused coping sub-scale and problem-focused sub-scale (Walter Reed Army Institute of Research, 1996). A secondary analysis was conducted on data collected from a survey study sample of 1,001 male Non-commissioned officers and enlisted soldiers deployed to Bosnia in March 1997. The results indicated that soldiers experiencing low levels of stress tended to perceive their leadership environment as positive and peer-relationships as cohesive. Furthermore, soldiers who used emotion-focused coping methods were likely to experience increasing levels of stress. Problem-focused coping had no statistically significant relationship to soldiers' stress levels. The findings are congruent with components of human relations theory (Follet, 1933; Barnard, 1938; Fayol, 1949) and transactional theory (Lazarus & Folkman, 1984).

Frank & Frank (1996) examined the unexplained illnesses that were documented in both veterans of the Persian Gulf War and in soldiers stationed in the Philippines during World War II. In both groups, patients exhibited similar symptoms (including weakness, fatigue, and headaches) that could not be attributed to any one source. Because troops in both settings also experienced intensely confusing and threatening information about their personal safety, they suggest that the symptoms represent physiological reactions to demoralizing stress.

Wright, Marlow & Gifford (1996) describe the stresses experienced by soldiers as they prepared for war during Operation Desert Shield, the buildup period to the Persian Gulf War. Information gleaned from interviews conducted during this tense period of uncertainty has provided important data on soldiers' adaptation, morale, cohesion, family (and personal) relationships, and concerns, as well as on potential problems they encountered. This includes observation of the effects of anticipatory stress and its ramifications for groups and individuals. The experiences and perceptions of combat and support units stationed in the Persian Gulf during the early months of the deployment were compared. Their work resulted in information that can present clinical, research, and community-based recommendations that can inform the actions of civic and military leaders, clinicians, and family members during future military contingencies.

Bartone & Ender (1994) reviewed how casualty policies have developed in the US Army, and draw on the Army's casualty experience to suggest some ways in which organizational responses to death might facilitate healthy adjustment for survivors. Military casualty activities serve important social and psychological functions because they impact on individual mental health and unit morale. A variety of studies (e.g., J.W.

Pennebaker et al., 1990) have shown that programs/activities that increase a sense of positive meaning regarding trauma and loss can facilitate healthy psychological adjustment for survivors. Casualty workers themselves can benefit from supportive organizational policies.

Armfield (1994) discussed various models for preventing posttraumatic stress disorder (PTSD) and examined future directions for PTSD prevention. Historically 10-50% of all casualties result in PTSD. The best treatment suggested by Armfield is rest and ventilation of feelings followed by return to duty and peer group. Preventing the PTSD cycle from starting and thus decreasing psychiatric casualties is reported as feasible. This can be done by promoting unit cohesion and morale, inducing stress during training so individuals will be better prepared to cope, providing realistic information about what to expect in combat, and holding group debriefings immediately after traumatic events. Stress inoculation therapy and critical incident stress debriefing are recommendations suggested.

Perceptions of Readiness

(Shamir, et al.,2000) conceptualized perceived combat readiness, an important component of morale, in terms of collective efficacy beliefs. They examined some of the anticipated correlates of collective efficacy beliefs as they apply to military combat units. They focused on the following variables: soldiers' experience, leader's tenure, leader's confidence in the unit, soldiers' confidence in the leader, unit discipline, and members' identification with the unit. The study was based on questionnaires given to company leaders (n=50), staff members (n=353), and two samples of soldiers (n=1,197) in 50 Israel Defense Forces companies. All participants were men (aged 18-23 yrs) Some support was found for all hypothesized correlates. The strongest predictor of perceived combat readiness was identification with the unit. The correlation between aggregated staff members' perceptions and aggregated soldiers' perceptions was only modest, suggesting that two groups may employ different standards to assess the combat readiness of their units.

West, Mercer & Altheimer (1993) examined the relationship of cohesion among members of a platoon leadership team to subordinate attitudes and performance, and the degree of consensus among squad members and their leaders regarding the leadership team's cohesiveness. Data from 1,281 military personnel from 60 light infantry platoons indicate that squad members' perceptions of leadership team cohesion

(LC) were significantly related (at both individual and unit levels) to perceptions of unit effectiveness. They were also related to ratings of performance in simulated combat. Regression revealed LC to be beneficial over and above the perceived quality of the individual leaders.

Slagle, et al., (1990) interviewed 37 crewmen involved in a nonfatal military aircraft accident, their spouses, fellow squadron members and their spouses, and individuals from the fire crew and control tower on duty during the accident. The age range was 18-45 years. Questionnaires assessed the presence of intrusive, avoidant, depressive/anxious, and posttraumatic stress symptoms. Spouses reported more symptoms after the accident than their flying husbands; intrusive symptoms were most common. Symptoms were correlated with various perceptions and experiences occurring before and following the accident. Several kinds of symptoms and experiences were significantly correlated with a perception of morale decline.

Preparing for Deployments

Franciskovic, Moro, Ljiljana & Palle-Rotar (1992) describe a 3-day program of psychological preparation for Croatian army soldiers and officers. Soldiers were taught simple techniques to overcome fear and tension in stress situations. Officers were taught to recognize group phenomena, and relations between a leader and the group were emphasized. The role of officers within the group and in the process of creating togetherness was addressed, and the importance of group communication was emphasized. The authors argue that work with officers is important because they are the most important factors in (1) forming cohesive groups of soldiers and (2) strengthening the military morale and efficiency of a brigade. A follow-up analysis of soldiers who had participated in training showed that strengthening unity within a group of soldiers helped to prevent undesirable psychical reactions to stress during the war.

Utilization of Mental Health Services

Constantian (1997) examined the need and use of outpatient mental health services by active duty members and active duty family members and to determine if the so-called "offset effect" could be detected in this population. Using a subset of the worldwide 1994-95 Department of Defense Health Beneficiary Survey numbering over 26,000 observations, several hypotheses were examined resulting in several important findings.

First, in spite of expectations of a more mentally fit active duty force, active duty members and family members have approximately the same mental health needs. Moreover, the aggregate mental health need (based on mental health status) was not statistically different from that of the general population. These findings defy the expectation that military members are more mentally fit due to entry level screening and programs designed to boost force morale.

Second, DoD beneficiaries exhibit lower levels of usage of mental health services than the general population. The active duty force's underutilization of mental health care is more marked than the underutilization attributable to active duty family members. Paradoxically, the Air Force, the service with the highest mental health status on average, had the highest mental health utilization rates.

The lack of consistency between need for and use of mental health services in the DoD requires further exploration. The underutilization of ambulatory mental health services suggests that programs for expanding access to care should be considered. Expanding access might result in a cost savings if the so-called "offset effect" was detected in the population, whereby users of mental health care with mental health problems generated fewer costs than non-users of such care because unnecessary physical health visits were reduced. However, no evidence of an offset effect was found in the aggregate or any subgroups examined. Instead, expansion of mental health access to those in need of such care, while perhaps beneficial from a quality of care and force treadiness perspective, is unlikely to be beneficial from a financial perspective. The absence of an offset effect means that remedying the mental health underutilization problem will not be accomplished easily or inexpensively.

Returning Veterans

West, Mercer & Altheimer (1993) describe the background, objectives, and services of a Department of Veterans Affairs Social Work Outreach Team's efforts to work with soldiers returning from Operation Desert Storm (ODS). The soldiers experienced discrimination, low morale, sexual harassment, and fear. Many experienced PTSD and needed crisis intervention. Interventions included individual sessions for support and insight; short-term psychotherapy; and a critical incident stress debriefing (CISD) model to assess and process traumatic events. The approach used by the outreach team gave social workers an opportunity to take the service to the veteran rather than to wait for the veteran to seek services or be referred. Early recognition of the traumatic experiences helps veterans to seek treatment sooner and to develop an awareness of symptoms of stress in themselves and family members.

Some Conclusions

Some of the important factors and variables related to morale that emerge in most of the above cited studies include: group leadership, health, levels of motivation, mental readiness, force structure, wellness behaviors, debriefings, social environment, work-family factors, family adjustment, unit cohesion, societal values, and others. These factors bear not only on the performance of the mi8litary member in the field, but also on his/her relationships with others, post deployment adjustment and family and societal interactions.

With deployments to the wars in Iraq and Afghanistan, issues of morale and unit cohesion will continue to be factors affecting re-adjustments to civilian life and return to equilibrium. Learning from past studies and experiences and providing needed and necessary support for family members and returning military members is crucial. Adequate uses of the debriefing process for military members and information and education sessions for families, friends and communities in general can help ease stressors, mitigate posttraumatic stress, and help with the needed adjustments to changes occurring in all involved.

References

Kohlberg, Lawrence; T. Lickona, ed. (1976). "Moral stages and moralization: The cognitive-developmental approach". *Moral Development and Behavior: Theory, Research and Social Issues.* Holt, NY: Rinehart and Winston.

Lazarus, R.S., & Folkman, S. (1984). *Stress, Appraisal and Coping.* New York: Springer

Chapter 3: Deployments and Warriors

Background

After over ten years of cumulative involvement in Iraq, Afghanistan and other related conflict situations, our National Guard, Reserve and Regular Military are returning to our communities around the country. Families, friends, children and Military members are becoming re-united. It is important that they are able to re-adjust and re-integrate back into the civilian world. This means that it is also important that communities, families and mental health/counseling professionals have an understanding of what to expect as this process occurs.

We have learned much over the years about what happens to our military members as they deal with the aftermath of war experiences. This book reviews some of what we have learned since the American Civil War in the mid-1800s through World War I, World War II, Korea, Viet Nam and various other conflicts, including Iraq and Afghanistan. It is important that, as mental health/counseling professionals helping and advising our military families and our communities, we have an understanding about what to expect and how to be able to deal with potential adjustments. These involve re-integrating into family, returning to a civilian work environment and handling the aftermath of exposures to war environments.

Ever since the Revolutionary War, the military in the United States has faced a variety of challenges as they have fought to protect and maintain freedoms and democracy for our citizens and of our allies and friends around the world. The Civil War, fought between North and South in the mid-1800s, was a divisive conflict that nearly tore the country apart. World War I was a major conflict during which the United States asserted itself for the first time as an international power to be reckoned with. We not only moved onto the world scene, changing the course of history, but also lost our innocence in the process. World War II saw the United States, together with her allies in Europe, the Pacific, Asia, Canada, and South America defeat ignorance, arrogance, prejudice, intolerance, despotism, tyranny, and dictators in a mammoth undertaking to free oppressed peoples and provide opportunities for more enlightened self-government and the promise of a better future for all. Many have referred to those who fought for this country in World War II as "The Greatest Generation". During the Korean War, the United States fought to maintain the promise of Democracy and self-determination on the Korean peninsula. This has never been resolved, but a divided Korea exists to this day under an Armistice that sees South Korea prospering. The Viet Nam War presented new challenges for our country and our military as we fought once more to help preserve hard-won freedoms and self-determination. This conflict was a very difficult, divisive and tough undertaking, resulting in Viet Nam making a determination of the form of government they want. While it may not be what many in the United States would have preferred, the two countries currently have cordial and positive relations with each other and look toward a future that promises to be prosperous and constructive for both.

Our country and our military have been involved in a number of less widespread, yet significant conflicts and missions to assist others in pursuing a future for all, free of tyrants and despots. The Cold War was pursued for many years with costs in money, people and ideas affecting many in our military who quietly fought for these ideals, yet were never totally acknowledged. Over the past 10+ years, our country and our military (including our National Guard and Reserves) have again been involved in a new and different kind of war to preserve, protect and defend our way of life and that of others.

Our military members in every one of these conflicts and wars have experienced many forms of wounds—physical, psychological, and emotional. In their efforts to accomplish their mission as military members, they have suffered much to help us all. Their families have been affected as well. As our military members return home following conflicts and multiple deployments that have gone on for over ten years, our civilian population struggles to assist their friends, family members, colleagues and neighbors to re-integrate back into civilian life. We have learned much over the years from history, experience and study about the effects of wars and conflicts on military and families. This can be useful in helping all concerned return to an equilibrium after such experiences. This book and set of essays attempts to review research, history and personal experiences in order to provide some educated approaches toward understanding the past and applying what we have learned toward planning for the future. It is obviously incomplete, yet hopefully assists in providing a beginning template and at least a general set of goals toward a positive and constructive future.

Our smaller and isolated rural and frontier communities are affected differently from suburban and more urban settings. Wyoming and other rural/frontier states in the western U.S. are sparsely populated (e.g., population of Wyoming is a little over 850,000), with a total of about 11,000 ranches and farms. Montana, rural Nevada, Idaho, Utah, New Mexico, the Dakotas, Alaska and other areas are in similar settings. The National Guard and Reservists from these areas have been involved in the Middle East, Latin America, the Caribbean, Afghanistan, and other conflict areas since the end of World War II. Stateside, they provide assistance during man-made and natural disasters, support our active military in various settings and are representative of our population. Many become members of the Guard and Reserve after high school and continue to remain active through college, work and other activities. Many first responders (e.g., nurses, firefighters, law enforcement, Civil Air Patrol, and others) remain actively involved and are deployed regularly into their 60s and beyond as members of the Guard and Reserve. Many of them grew up on ranches, farms and other rural settings.

Western Rural Demographics

Cell phone systems in many areas of the rural west regularly experience "black holes" with no service. Computer systems in many rural/frontier settings are only beginning to develop the uses of computers at full potential. Computers are used regularly in many schools (more so than ten years ago). Most county fairs focus strongly on rural activities. All have a strong understanding and appreciation of rural values. These may be similar in some ways to those that existed in most farming and ranching areas throughout the country in the 1950s. They remain quite vibrant in the rural west and are alive and well through many rural organizations such as Farm Bureau. 4H, FFA, Rodeo, Ag Extension Services and other related agricultural organizations. Many western rural residents are descendants of ranchers and farmers who originally homesteaded their lands. People in the west are quite often members of the Guard and Reserve. Some also become active duty members of the Navy, Air Force or Marines. Agriculture and Energy development (e.g., wind, solar, nuclear, etc) are two of the major industries in the rural west. Agriculture (supported by a number of major Agriculture Departments in Land Grant Universities) provides a major proportion of the national food supply. Energy development (e.g., coal, oil, uranium, wind, solar and others) are increasingly important in western states.

This book provides a background of previous work, research and experiences that can help our mental health/counseling professionals and communities better prepare to be positive resources for our civilian military members, families and our communities. National Guard and Reserves (both Army and Air Force) are integral parts of the rural west and provide important service at home. In recent years as in the past, our civilian military have also provided service to the country in international settings.

PART III:
Stress and Stress Responses

Chapter 4: Delayed Stress

"Four things come not back: the spoken word, the spent arrow, the past, the neglected opportunity."

Omar Idn Al-Halif

There have been some dramatic events in the news over the past few years. Tragedies such as earthquakes, floods, hurricanes, tornadoes, bombings, etc. While certainly not new or even uncommon, they are beyond our abilities to control or, in some cases, even difficult to fully comprehend. In Oklahoma City, rescue workers tried to help survivors and non-survivors following the bombing of a Federal building. We all have seen scenes on the evening news and elsewhere of the aftermaths of the bombings of the U.S. Embassies in Nairobi and Dar Es Salaam, Baghdad, Afghanistan, and the tragedy of September 11 in New York. Following their return from the Viet Nam War, many Veterans experienced PTSD. Veterans of the Gulf War seem to be dealing with a similar problem, which has been called Gulf War Syndrome.

What do all of the above have in common? Studies of PTSD and how to treat it and/or how to prevent it and many of its symptoms was the result of a lot of work with Veterans. Today, we know more about PTSD and how to alert those at risk (e.g., rescue workers, victims of physical and sexual abuse, victims of natural and man-made disasters) so that they can be better prepared to deal with some of the inevitable feelings and thoughts as well as other symptoms. PTSD, sometimes also referred to as Delayed Stress, is identified by certain common signs, including the following:

Depression

Depression is a common response to a traumatic event. It can be present in a number of forms, which may include sleep disturbance (e.g., difficulties sleeping, intrusive or disturbing dreams, or even too much sleep). Other signs may include general feelings of worthlessness or helplessness or difficulties concentrating. Some may experience feelings that no one will understand how they feel. They may find little support among friends, acquaintances and/or relatives. Some may try to alleviate their feelings through attempts at "self-medication" involving alcohol and/or drugs.

Isolation

There are times when those involved isolate themselves from others or will have few friends. They may feel isolated and distant from peers. For example, they may feel that their peers or friends and family would rather not hear what their experiences were like. They may feel rejected.

Rage

Rage is also a common response. It involves feelings of irritation, touchiness, easily striking out at others who happen to be near (usually verbally, but sometimes physically). Some may experience frequent rage reactions while others may sublimate or repress their rage by breaking inanimate objects or putting fists through walls. There are many reasons for the rage—a rage at not being able to control or change the events that had occurred.

Avoidance of Feelings

Some may talk about episodes in which they did not feel anything when they witnessed or experienced the death of a buddy or friend or the more recent death of a close family member. Often troubled by their responses to tragedy, on the whole, they would rather deal with tragedy in their own detached way. Especially problematic is an inability to experience the joys of life. They may describe themselves as being emotionally dead. This "defense mechanism" dulls one's awareness of the death and/or destruction surrounding him/her. It is a survival mechanism which does help one to pass through a period of trauma without becoming caught up in its tendrils. This numbing only becomes nonproductive when the period of trauma has passed, and the individual is still numb to the affect or emotions around him/her. They may feel that, should they let themselves release the numbness, they may never stop crying or may completely lose control of themselves.

Survival Guilt

When others have died and some have not or are rescuers, they may ask, "How is it that I survived when others more worthy than I did not?" or "What could I have done to get here sooner and save this life?" Survival guilt is an especially guilt-provoking symptom. It is not based on anything hypothetical. Rather, it is based on the harshest of realities—the actual death of a human being and the struggle of the survivor or rescuer to live. In some cases, the survivor or rescuer has had to compromise him/herself or the life of someone else in order to deal with this. The guilt that results may eventually lead to self-destructive behaviors. Feelings of helplessness may develop over the inability to change the outcome of events. Guilt may develop over "maybe if I had been there sooner or had done more, etc..." In some cases (e.g., war, earthquakes, other disasters) those who suffer the most painful symptoms are primarily those who have served as corpsmen, medics, EMTs, etc. They save many lives. However, some of those they try to save die. Many casualties are beyond medical help, yet many emergency response workers suffer extremely painful memories for long periods thereafter—some for the remainder of their lives. Some tend to blame themselves for these deaths.

Anxiety Reactions

Many workers describe themselves as very vigilant human beings. Their autonomic senses are tuned to anything out of the ordinary. Fear is a normal reaction to disaster, frequently expressed through continuing anxieties about recurrence of the disaster, injury, death, separation and loss.

Intrusive Thoughts

Some workers frequently report replaying especially problematic experiences over and over again. They may search for alternatives to what actually happened. They may castigate themselves for what they might have done to change the situation, suffering subsequent guilt feelings today because they were unable to do so in the situation. Most report that these thoughts are very uncomfortable, yet they are unable to put them to rest. Not all who are exposed to tragedies experience all or some symptoms. However, it is not possible to be exposed to such events and have no feelings, thoughts or responses. Being prepared for them and recognizing what effects they can have can prepare one to deal with them. Debriefing following involvement in such experiences can help prepare one for what they may feel or think. It can provide them with support from others; methods for dealing with their feelings, thoughts and responses; and resources for future assistance if needed. It is like a vaccination which helps prevent more serious consequences.

Chapter 5: PTSD and Suicide Following War

"The best mirror is a friend's eye."

Gaelic Proverb

When one or more members of a family are traumatized, the entire family can suffer from posttraumatic symptoms. Unfortunately, this may go unrecognized by the family, friends, and professionals. A cycle of post-traumatic victimization and fragmentation of family integrity can lead to disastrous consequences. The posttraumatic phases leading to such a destructive outcome can potentially involve events like a young adult child's suicide, combat trauma and loss, or a child's witness of parental suicide. Treatment of traumatized families may include psychoeducational, psychodynamic, systemic, behavioral, and spiritual interventions.

Trauma has a big impact on both individuals and society as a whole. According to Davidson (2000), recognition of the extent of this impact by the medical profession has been relatively slow. However, with growing appreciation of the prevalence of trauma exposure in civilian as well as combat populations, the true scale of trauma-related psychiatric consequences is beginning to emerge. Reports cited in Davidson suggest that more than 60% of men and 51% of women experience severe psychiatric stress that is compounded by significant comorbid illness. This impacts critically upon quality of life, resulting in grave functional and emotional impairment. Additionally, there is a detrimental cost to society with high financial and social consequences from the significantly elevated rates of hospitalization, suicide attempts and alcohol abuse (Davidson, 2000), Davidson et al. (1991) examined PTSD among 2,985 people (aged 18-95 years). Based on responses to the Diagnostic Interview Schedule, the lifetime and six month prevalence figures for PTSD were 1.3% and 0.44%, respectively. Compared with non-PTSD respondents, those with PTSD had significantly greater job instability, family history of psychiatric illness, parental poverty, experience of child abuse, and parental separation or divorce before age 10 years. PTSD was associated with greater psychiatric comorbidity, attempted suicide, social phobia, obsessive-compulsive disorder, generalized anxiety, and major depression. Compared with acute cases, chronic PTSD was accompanied by more social phobia and somatization disorder, impairment of subjective social support, and greater likelihood of physical attack with regard to initiating trauma.

Amir et al., (1999) examined the relationship between suicide risk and coping styles in 46 PTSD sufferers (mean age 39 yrs) as compared with 42 patients diagnosed with other anxiety disorders (mean age 42 yrs) and 50 healthy controls. Subjects (S) completed the Suicide Risk Scale (SRS) and the Albert Einstein College of Medicine Coping Styles Questionnaire based on the R. Plutchil (1991) model of emotions. The PTSD Ss scored significantly higher than the two control groups on the SRS. In the PTSD group, suicide risk was significantly negatively correlated with the coping styles of mapping, minimization, and replacement and positively correlated with the coping style of suppression. Coping styles significantly explained the variance in the SRS scores for all three groups. The cognitive map of PTSD patients highly resembles other populations with high suicide risk. Amir et al's (1999) results suggest that clinicians treating victims of traumatic events should focus on problem-solving therapies in order to help these patients deal less rigidly with everyday stresses and decrease suicide risk.

Ferrada-Noli et al., (1998) assessed the prevalence of PTSD and psychiatric comorbidity, the incidence of suicidal behavior among refugees with a history of exposure to severe trauma, and the possible difference between the diagnoses with respect to modes of suicidal behavior. 149 adult refugees with severe traumatic experiences underwent PTSD diagnoses and an assessment of suicidal behavior. PTSD prevalence was 83% in all cases in which a principal psychiatric diagnosis was established. A significant over-representation of suicidal behavior was found in Ss with PTSD compared with non-PTSD Ss. No difference was found with respect to the total prevalence of suicidal behavior between depressed and non-depressed PTSD subgroups. Non-depressed PTSD patients showed an increased frequency of suicide attempts, but decreased frequency of suicide thoughts, relative to depressed PTSD patients.

Ferrada-Noli, Asberg & Ormstad (1998) studied whether relationships exist between the type of torture stressors and suicidal ideation. The hypothesis was that the nature of the torture methods would be reflected in the content of posttraumatic self-destructive ideation. 65 adult refugees who had been assessed with both diagnoses of PTSD and suicidal behavior underwent psychiatric diagnoses and suicidal behavior assessment. Results showed a clear association between the mode of torture and preferred suicidal strategy. Among PTSD patients with a history of torture, an association was seen between the torture methods that the victim had been exposed to, and the suicide method used in ideation or attempts. Blunt force applied to the head and body was associated with jumping from a height or in front of trains, water torture with drowning, or sharp force torture with methods involving self-inflicted stabbing or cutting. These suggested relationships between main stressors and content of suicidal ideation.

Vietnam Veterans

In our current decade, the Vietnam War still haunts the American conscience, not only because of the 57,939 Americans who lost their lives, but also due to a much larger number (up to 1.5 million by some estimates), who returned with war-induced Post-Traumatic Stress Disorder (PTSD), a delayed stress syndrome responsible for a wide array of psychological and social problems. However, despite the centrality of the Vietnam Veteran to American views on war and despite the oft-repeated claim that the psychological problems of the Vietnam Veteran are unique in American history, there has been no serious attempt to place these problems into any meaningful historical context. Nor, despite over 50,000 books having been written on the subject over the past 130 years, has there yet been any attempt to investigate the psychological and related readjustment problems of veterans of the American Civil War. Dean (1996) attempts to place the problems of Vietnam Veterans into a meaningful historical context by investigating the psychological problems of soldiers and veterans of the American Civil War (1861-1865). Insane asylum and federal pension records bearing on the readjustment problems of Civil War veterans were closely analyzed, and reveal that Civil War soldiers and veterans experienced a wide array of mental problems, from depression, anxiety, flashbacks, and cognitive disorders (loss of concentration and memory), to resulting social pathologies such as suicide, alcoholism, and domestic violence. Hence, the Vietnam Veteran was not unique in American history in this respect. However, Dean also rejects as simplistic the post-Vietnam tendency to view all American veterans as neglected and badly treated. Rather, he suggests that war is a complex phenomenon. While some soldiers are devastated by the attendant hardships and danger, others are strengthened by the experience. Additionally, views of war and the veteran are culturally conditioned.

Employing a multidisciplinary approach that merges military, medical and social history, Dean (1997) draws on individual case analyses and quantitative methods to trace the reactions of Civil War veterans to combat and death. He seeks to determine whether exuberant parades in the North and sectional adulation in the South helped to wash away memories of violence for Civil War veterans. His study reveals and supports his previous study that Civil War veterans experienced severe persistent psychological problems such as depression, anxiety and flashbacks, with resulting behaviors such as suicide, alcoholism, and domestic violence. By comparing Civil War and Vietnam Veterans, Dean demonstrates that Vietnam Vets did not suffer exceptionally in the number and degree of their psychiatric illnesses. The politics and culture of the times, he argues, were responsible for the claims of singularity for the suffering Vietnam Veterans as well as for the development of the modern concept of PTSD.

Faberow, Kang & Bullman (1990) examined potential risk factors for suicide from records of 175 Vietnam Veterans (VVs) or non-VVs who committed suicide or died from motor vehicle accidents MVAs). No military service factor was associated with suicide. Characteristics of VV suicides were not substantially different from non-VV suicides with respect to known demographic risk factors. The psychological profiles of VV suicides were also similar to non-VV suicides in most instances. Symptoms related to PTSD were observed more frequently among suicide than MVA cases, However, suicides were not associated with specific combat experiences or military occupation. The extent of combat experience in Vietnam was not a good predictor of suicide death.

Fontana, Rosenheck & Brett (1992) extended the understanding of war zone stressors by specifying the psychological meaning of traumas anf by examining the extent to which this specification adds to the ability to account for the severity of current symptoms of PTSD. Eleven traumas were organized in terms of four roles that veterans played in the initiation of death and injury; namely, target, observer, agent, and failure. The relationships of these roles to current symptomatology were examined in combination with a set of objective measures of war zone stressors. Ss were 1,709 Vietnam theater Veterans. Results suggest that having been a target of others' attempts to kill or injure was related more strongly than other roles to symptoms of PTSD. On the other hand, having been an agent of killing and having been a failure at preventing death and injury were related more strongly than other roles to general psychiatric distress and suicide attempts.

Fontana & Rosenheck (1995) investigated the etiology of attempted suicide using both retrospective and prospective data from 402 Vietnam Veterans who were receiving treatment in the Department of Veterans Affairs Posttraumatic Stress Disorder Clinical Teams Program. The applicability of a community-based model to the treatment-seeking sample was assessed. The community-based model achieved a very high fit with reasonably good parsimony with the treatment-seeking data. Causal paths in the treatment-seeking sample mirrored those in the community sample in that psychiatric symptoms (including PTSD) were the sole factors contributing directly to attempted suicide. Traumatic military experiences played a substantial role, but only indirectly as they contributed to the development of psychiatric symptoms.

Data from the National Vietnam Veterans Readjustment Study were used to develop an etiological model of attempted suicide among a community sample of 1,198 male Vietnam Veterans. In a 30-step process, Fontana & Rosenheck (1995) used structural equation modeling to develop a model that they refined, cross-validated, and then specified in terms of its replicable paths. The final model possessed highly satisfactory fit and parsimony. General psychiatric disorders (GPDs) were the sole factors contributing directly to attempted suicide. GPDs were in part products of both non-military and military traumas, most specifically participation in abusive violence. Substance abuse and PTSD were related to attempted suicide bivariately but not when considered in conjunction with GPDs. Among pre-military risk factors, family instability contributed to attempted suicide indirectly through its influence on GPDs.

Freeman et al., (1995) hypothesized that suicide attempts in 14 combat veterans with chronic PTSD represent impulsive, aggressive, and self-destructive acts, and therefore they would exhibit a more frequent history of other aggressive and impulsive acts relative to their 16 peers who had not made suicide attempts in the past. Findings did not support the hypothesis. The two groups appeared to engage in impulsive, violent behaviors with weapons at a similar frequency. There were no significant differences between the groups in terms of reported PTSD symptom severity and histories of alcohol abuse, substance abuse, and combat exposure. The only substantial difference between groups lay in their activities, drug/alcohol problems, or suicide attempts. Low childhood adjustment ratings and school suspensions predicted adult alcohol abuse, respectively (Hiley et al., 1995).

In summary, the above studies identify a number of variables that should be considered when reviewing the adjustment of Iraq War combat and Afghanistan veterans when they return to civilian life or rotate assignments back. These include the extent and types of combat experiences. Combat guilt was found in a number of previous studies to be a strong predictor of suicide attempts and preoccupation with suicide. However, the extent of combat experience in at least one study was found to not be a good predictor of suicide. Problem areas identified which should be assessed for in returning military include depression, anger, guilt about combat actions, survivor guilt, anxiety, domestic conflicts, substance abuse, traumatic memories, and others. It is likely that those who were exposed to more severe circumstances (e.g., losing a close buddy in combat or as a prisoner) will be more at risk. Individual levels of resilience are likely to vary for a variety of reasons.

References

Amir, Marianne, Kaplan, Z., Efroni, R. & Kotler, M. (Mar-Apr 1999). Suicide risk and coping styles in posttraumatic stress disorder patients. *Psychotherapy & Psychosomatics, Vol 68(2), pp. 76-81.* Journal URL: http://www.karger.ch/journals/pps_jh.htm

Brende, Joel O. & Goldsmith, Richard (Sum 1991). Post-traumatic stress disorder in families. *Journal of Contemporary Psychotherapy,* Vol 21(2), pp. 115-124. Journal URL: http://www.wkap.nl/journalhome.htm/0022-0116

Brent, David A., Moritz, Grace, Bridge, Jeff, Perper, Joshua et al. (May 1996). Long-term impact of exposure to suicide: A three-year controlled follow-up. *Journal of the American Academy of Child & Adolescent Psychiatry,* Vol 35(5), pp. 646-653. Journal URL: http://www.jaacap.com/

Brent, David A., Perper, Joshua A., Moritz, Grace, Liotus, Laura et al.

(Feb 1995). Posttraumatic stress disorder in peers of adolescent suicide victims: Predisposing factors and phenomenology. *Journal of the American Academy of Child & Adolescent Psychiatry*, Vol 34(2), pp. 209-215. Journal URL: http://www.jaacap.com/

Bryant, Richard A. (Jul 1998). An analysis of calls to a Vietnam veterans' telephone counseling service. *Journal of Traumatic Stress*, Vol 11(3), pp. 589-596. Journal URL: http://www.wkap.nl/journalhome.htm/0894-9867

Bullman, Tim A. & Kang, Han K. (1997). Posttraumatic stress disorder and the risk of traumatic deaths among Vietnam veterans. In Fullerton, Carol S. (Ed); Ursano, Robert J. (Ed); *Posttraumatic stress disorder: Acute and long-term responses to trauma and disaster*. Progress in Psychiatry series, No. 51. pp. 175-189.

Bullman, Tim A. & Kang, Han K. (Nov 1994). Posttraumatic stress disorder and the risk of traumatic deaths among Vietnam veterans, *Journal of Nervous & Mental Disease*, Vol 182(11)[1344], pp. 604-610. Journal URL: http://www.jonmd.com/

Carter, Bonnie Frank & Brooks, Allan (1991). Child and adolescent survivors of suicide. In Leenaars, Antoon A. (Ed); *Life span perspectives of suicide: Time-lines in the suicide attempts. Journal of Nervous & Mental Disease*, Vol 183(10), pp. 664-666. Journal URL: http://www.jonmd.com/

Goodale, Peggy Ann (April 1999). Anger profile of suicidal inpatient Vietnam veterans. (posttraumaticstress disorder). *Dissertation Abstracts International: Section B: The Sciences & Engineering*, Vol 59(18-B), 5577.

Hendin, Herbert (Jan 1992). "PTSD and risk of suicide": Reply. *American Journal of Psychiatry*, Vol 149(1), pp. 143. Journal URL: http://ajp.psychiatryonline.org/

Hendin, Herbert & Haas, Ann P. (May 1991). Suicide and guilt as manifestations of PTSD in Vietnam combat veterans. *American Journal of Psychiatry*, Vol 148(5), pp. 586-591. Journal URL: http://ajp.psychiatryonline.org/

Hiley-Young, Bruce, Blake, Dudley David, Abueg, Francis R., Rozynko, Vitali et al (Jan 1995). Warzoneviolence in Vietnam: An examination of premilitary, military, and postmilitary factors in PTSD in-patients. *Journal of Traumatic Stress*, Vol 8(1), pp. 125-

141. Journal URL: http://wwwwkap.nl/journalhome.htm/0894-9867

Plutchik, Robert (1991). *The Emotions, Revised Edition.* University Press of America.

Recommended Reading

Dean, E.T. (1997). Shook over hell: Post-traumatic stress, Vietnam, and the Civil War. Cambridge Mass: Harvard University Press

Additional Readings: Search Terrorism and Stress in the search engine. Also try looking for Psychology and Terrorism.

Chapter 6: Fear and Anxiety

War and Anxiety

A war or a national crisis situation is accompanied by an increase in stress and anxiety levels within the population—among civilians as well as military. It also results in posttraumatic stress in both populations which can last for very extended periods of time. For example, Lindorff (2002) notes that little is known about the psychological effects of war service on Australian World War II veterans. In an attempt to understand some of these effects, 88 survivors (aged 75-91 yrs) of one of the war's most intense actions (the Battle of Isurava on the Kokoda Track in Papua in August 1942) responded to a survey asking for their recollections of the battle, and for a description of its effect on them. Many said that they had yet to recover from the experience. Large numbers indicated continuing ill effects. These included nightmares, sleeplessness, negative imagery, "flashbacks", problems with concentration, weeping, generalized anxiety, and distress caused by situations recalling the battle. Many commented that they had never talked to anyone about their war experiences, or the effects of these experiences. Only two veterans reported seeking or receiving any treatment for their symptoms (Lindorff, 2002).

Traumatic experiences associated with the war in Bosnia (1992-1995) impacted the lives of many Bosnian refugees and displaced people. Approximately 25% of Bosnians were forced to leave their homes and resettle in other areas of Bosnia or abroad. Plante et al (2002) describe war-related stress and the association of marital status, anxiety, depression, and sensitivity levels. 82 displaced Bosnians living in the area of Tuzla, Bosnia, and 53 refugees living in the San Francisco Bay area in the U.S. completed the same questionnaire in the Bosnian language. Better self-reported health was related to being single, having lower anxiety ratings, finding and adapting to a new environment easily, and moving on with life. Findings also revealed that being divorced or separated, better self-reported health, and lower anxiety, depression, and sensitivity ratings were predictors of more effective coping.

Thabet and Abed (May 2002) assessed the nature and severity of emotional problems in 91 Palestinian children (aged 9-18 yrs) exposed to home bombardment and demolition during Al Aqsa Intifada, and 89 age-matched controls, echo completed self-report measures of post-traumatic stress (PTS), anxiety, and fear. Significantly more children exposed to bombardment and home demolition reported symptoms of PTS and fear than controls. 54 (59%) of 91 exposed children and 22 (25%) of 89 controls reported PTS reactions of clinical importance. Exposure to bombardment was the strongest predictor of PTS reactions. By contrast, children exposed to other events, mainly through the media and adults, reported more anticipatory anxiety and cognitive expressions of distress than children who were directly exposed. Children living in war zones can express acute distress from various traumatic events through emotional problems that are not usually recognized. Health professionals and other agencies coming in contact with children who have been affected by war and political violence need to be trained in detection and treatment of such presentations.

As part of a United Nations Children's Fund's (UNICEF) psychosocial program during the war in Bosnia-Herzegovina, data were collected from a community sample of 2,976 children, aged between 9 and 14 years (Smith et al, Apr 2002). Children standardized self-report measures of posttraumatic stress symptoms, depression, anxiety, and grief, as well as a report of the amount of their own exposure to war-related violence. Results showed that children reported high levels of posttraumatic stress symptoms and grief reactions. However, their self-reported levels of depression and anxiety were not raised. Levels of distress were related to children's amount and type of exposure. Girls reported more distress than boys but there were few meaningful age effects within the age band studied. Even a passing interest in daily news events confirms that violent crime, disasters, serious accidents, war, and other forms of traumatic events occur frequently and with great impact upon individuals, families, and communities. Research studies have confirmed several crucial facts regarding the nature of traumatic events (Freedy & Hobfol, 1995). These facts include:

- That traumatic events occur frequently, impacting large

numbers of people;

- That exposure to traumatic events substantially increases the risk of several serious mental health problems; and

- That it is possible to limit the impact of traumatic events through the application of prevention, assessment, and treatment strategies.

The experience of wartime stress may change a certain aspect of an individual's personality, in particular the personality trait of neuroticism defined as "proneness to distressing emotional states" such as anxiety, depression and anger. Bramsen et al (2002) examined this by studying a random community sample of 455 Dutch survivors (aged 63-72 yrs) of World War II. The relationship between wartime stress and the personality trait of neuroticism turned out to be fully mediated by the development of a negative worldview. Empirical support was found for the notion that traumatic events force the survivor to change the personal theory of the world and make him/her more vulnerable to distressing emotional states and symptoms of posttraumatic stress disorder.

Is There a Fear of Other Cultures?

Fears can sometimes be transferred from older sources to newer or different ones once the original has been eliminated. Murray & Meyers (1999) used the collapse of the Soviet Union to test the hypothesis that some people are psychologically predisposed to "need enemies". The findings from the 1988-1992 Leadership Opinion Project (LOP) panel data show that those respondents who had been highly suspicious of Soviet motives before the end of the cold war were more likely to view other countries with suspicion and to perceive the international environment as dangerous after the Soviet collapse. There is no evidence that people have actually transferred old fears about the Soviet Union onto a replacement enemy.

China is the country most frequently named as the U.S.'s main adversary following the cold war, making it the most likely object for the transference of hostility. However, even the most ardently anti-Soviet respondents failed to exhibit greater fear or animosity toward China after having lost their old enemy. Nonetheless, it is possible that the need to have some sort of "enemy" was transferred to Al Qaeda and Saddam Husein's Iraqi regime.

Culturally Relevant Approaches toward Healing

Culturally relevant approaches toward healing are important considerations when developing and implementing interventions within a culture not one's own. For example, among Somalis in Ethiopia, war-related distress is not interpreted in a medical framework aimed at healing. Rather, such violence is predominantly assimilated into the framework of Somali politics, in which individual injuries are considered injuries to a lineage or other defined group. The dominant emotion in this context is not sadness or fear, but anger, which has emotional, political and material importance in validating individuals as members of a group sharing mutual rights and obligations (Zarowsky, 2000). Before advocating trauma-based models of war-related distress, researchers and practitioners should consider whether a medical framework would do better at helping individuals and communities to deal with distress and reconstruct meaningful lives and relationships in circumstances of long-standing collective violence. It is important to consider culturally relevant approaches before proceeding in any response. For example, Hinterhuber et al (2001) investigated attitudes toward warfare in the former Yugoslavia among refugees and emigrants. 283 Serbs, Montenegrins, Croats, and Bosnians (mean ages 32 - 38 yrs) completed surveys concerning:

- feelings, values, political attitudes, and assessments of contemporary history;
- attitudes to the then ongoing war;
- prejudices; and
- perspectives for the future.

Results show that almost half of Croat respondents were happy when Yugoslavia started to disintegrate. The Serbs predominantly felt grief, followed by fear and uncertainty.

Dar, et al (1999) studied the effects of military service during the Intifada on veterans from a kibbutz background. 184 Israeli Defense Forces veterans (133 males and 51 females) from a kibbutz background who had served in the occupied territories during the Intifada (1992-1998) were retrospectively asked in a semi-structured questionnaire how this service had affected them. The results showed some common themes:

- service in the "territories" deepened the person's understanding of the Israeli-Palestinian conflict, but also increased their fear and hatred of Arabs;

- few respondents developed a new political position but were firmer in their original "leftist" attitudes;
- only a few based their position on support for the Palestinians' rights or suffering, focusing mainly on a generally utilitarian consideration of the Israeli side's needs;
- in order to cope with the conflict between their military duty and the internalized values of their kibbutz education, they either sought shelter behind army orders or compartmentalized their humanistic values and military duty; and
- they regarded their military service during the Intifada as a most difficult experience but leaving only a situational, temporal psychological imprint.

Cultural values, exposure to traumatic events, civilian or military status, attitudes, history, prejudices, politics, education, and other variables all play a role in the experience of and effective responses to fears due to war situations.

Fear and Nuclear Threats

Not since 1945 has the world experienced nuclear warfare, although there has been the threat of nuclear terrorism and a large number of nuclear/ radiological accidents. Most people fear a nuclear/radiological threat even more than a conventional explosion due to their inability to perceive the presence of radiation with the ordinary human senses and to concerns about perceived long-lasting radiation effects. Studies of radiological accidents have found that for every actually contaminated casualty, there may be as many as 500 people who are concerned, eager to be screened for contamination, sometimes panicked, and showing psychosomatic reactions mimicking actual radiation effects (Salter, 2001). Data from the Hiroshima and Nagasaki attacks revealed widespread acute reactions such as psychic numbing, severe anxiety, and disorganized behavior, and later there were chronic effects such as survivor guilt and psychosomatic reactions. Such responses would likely be common in any future nuclear/radiological accident, terrorist attack, or warfare.

The response to 'new security' risks requires significant changes in public behavior, and the legitimization of unpopular government policies. Public education is one means of achieving this. The need is reflected in initiatives such as environmental and developmental education, health promotion, and the public understanding of science. Current strategies are often based on commercial advertising, but mass communications theory does not directly encompass influencing perception, which is necessary to create awareness of the new 'invisible' risks. Recent evolutionary brain science is providing new insights into our species perception deficits, which can inform a more effective approach to public education. Williams (2002) places risk within the context of the post-Cold War 'global security' agenda. He proposes fundamental areas of evolutionary perception: fear and disgust, number perception, and cheating. This leads to a core concept for public education about new security risks, 'enhanced difference', and a set of hypotheses that can be applied to text or image. Whatever approach, fear remains a potent motivator of behavior. It is one that is consistently utilized for control and is manipulated by others for maximum effects. Helping those traumatized by fears is a major task for psychologists and others attempting to help them readjust to the world and to achieve a level of equilibrium.

The potential for war is a pervasive threat to the security and family structure of children in military families. Ryan-Wenger (2001) compared 91 children (8-11 yrs old) of active-duty, reserve, and civilian families with respect to their perceptions of war, origin of fears related to war, levels of manifest anxiety, coping strategies, and projection of emotional problems in human figure drawings. Her findings regarding the adaptation of children in military families suggest the need for further research from children's perspectives is important for understanding adaptations and responses of military children.

Major factors affecting families with military members include war-zone military service, family separation, and readjustment back into the community by service members, posttraumatic stress (including PTSD) and psychosocial malfunctioning are among problems encountered. Strengths that contribute to resiliency by all family members include religious values, a positive outlook on life events, family support and various forms of psycho-social interventions. Children living and surviving in war zones are affected adversely in a number of developmental ways and are at severely increased risk of becoming unproductive members of their society.

References

Dar, Y., Kimhi, S., Stadler, N. & Epstein, A. (May 1999). Imprint of the Intifada: Response of kibbutz born veterans to military service in the West Bank and Gaza. *Megamot*, Vol 39(4), pp. 420-444. Publisher URL: http://www.szold.org

il de Silva, H., Hobbs, C. & Hanks, H. (2001). Conscription of children in armed conflict–a form of child abuse. A study of 19 former child soldiers. *Child Abuse Review*, Vol 10(2), pp. 125-134. Journal URL: http://www.interscience.wiley.com/jpages/0952-9136/

Havenaar, J.M., (Ed), Cwikel, J.G., (Ed) & Bromet, E.J., (Ed) (2002). *Toxic turmoil: Psychological and societal consequences of ecological disasters*. Series Title: Plenum series on stress and coping. New York, NY, US: Kluwer Academic/Plenum Publishers. xiii, 279 pp.

Herman, P. (2000). Children and the trauma of war: Exploring the use of games in transforming attitudes and behaviors. *Dissertation Abstracts International: Section B: The Sciences & Engineering*, Vol 61(3-B), pp. 1637.

Hinterhuber, H., Stern, M., Ross, T.s & Kemmler, G. (Oct 2001). The tragedy of wars in former Yugoslavia seen through the eyes of refugees and emigrants. *Psychiatria Danubina, Vol 13(1-4)*, pp. 3-14.

Kopelman, M. D. (Oct 2000). Fear can interrupt the continuum of memory. Journal of Neurology, *Neurosurgery & Psychiatry*, Vol 69(4), pp. 431-432. Journal URL: http://jnnp.bmjjournals.com/

Murphy, Ronald T., Wismar, Keith & Freeman, Kassie (Feb 2003). Stress symptoms among African-American college students after the September 11, 2001 terrorist attacks. *Journal of Nervous & Mental Disease*, Vol 191(2), pp. 108-114. Journal URL: http://www.jonmd.com/

Murray, Shoon Kathleen & Meyers, Jason (Oct 1999). Do people need foreign enemies? American leaders' beliefs after the Soviet demise. *Journal of Conflict Resolution*, Vol 43(5), pp. 555-569. Publisher URL: http://www.sagepub.com

Salter, C.A. (Dec 2001). Psychological effects of nuclear and radiological warfare. *Military Medicine*, Vol 166(12,Suppl 2), pp. 17-18. Publisher URL: http://www.amsus.org/

Starcevic, V., Kolar, D., Latas, M., Bogojevic, G. & Kelin, K. (2002). Panic disorder patients at the time of air strikes. *Depression &*

Anxiety, Vol 16(4), pp. 152-156. Publisher URL: http://www.interscience.wiley.com

Thabet, Abdel Aziz Mousa, Abed, Yehia & Vostanis, Panos (May 2002). Emotional problems in Palestinian children living in a war zone: A cross-sectional study. *Lancet*, Vol 359(9320), pp. 1801-1804.

Valkenburg, P.M., Cantor, J. & Peeters, A.L. (Feb 2000). Fright reactions to television: A child survey. *Communication Research*, Vol 27(1), pp. 82-97.

Williams, C. (Jul 2002). 'New security' risks and public educating: The significance of recent evolutionary brain science. *Journal of Risk Research*, Vol 5(3), pp. 225-248. Publisher URL: http://www.tandf.co.uk

Witty, C.J. (Oct 2002). The therapeutic potential of narrative therapy in conflict transformation. *Journal of Systemic Therapies*, Vol 21(3), Special Issue: Reflections in the aftermath of September 11. pp. 48-59. Journal URL: http://www.guilford.com/cartscript.cgi?page=periodicals/jnst.htm&cart_id=547216.21319

Zarowsky, C. (Sep 2000). Trauma stories: Violence, emotion and politics in Somali Ethiopia. *Transcultural Psychiatry*, Vol 37(3), pp. 383-402.

Recommended Reading:

Ross, G. (2003). *Beyond the trauma vortex: The media's role in healing fear, terror, and violence.* Berkeley, Calif: North Atlantic Books
Additional Readings: Search Terrorism and Stress in a search engine. Also try looking for Psychology and Terrorism.

PART IV:
Returning To Equilibrium

Chapter 7: Disequilibrium?

"When people speak to you about a preventive war, you tell them to go and fight it. After my experience, I have come to hate war. War settles nothing."

—Dwight D. Eisenhower

National traumas in recent years (e.g., 9/11/2001; other terrorist acts and threats; military actions in Afghanistan and Iraq; natural disasters such as Katrina and Rita, etc.) have heightened interest in the mental health needs of persons defending national interests, participating in peacekeeping missions, and maintaining a civil society are a critical part of strengthening and maintaining our national infrastructure.

Adjustment disorders can cause a variety of problems for individuals affected, including psychological pain and suffering, and premature mortality. These problems can extend to spouses and children in the form of child and family adjustment disorders. Due to their occupational roles in society, certain groups (e.g., first responders, military) are at heightened risk for trauma exposure and associated health and functional sequelae. Similar to the general population of trauma survivors, risk and adverse adjustment are not equally distributed among emergency responders and other high-risk occupational groups (e.g., military members.

Much research has focused on a variety of approaches to help improve our ability to deal with the results of traumatic exposure. Many of these include early interventions to reduce the likelihood of chronic posttraumatic stress. Other research has tried to identify biomarkers of vulnerability and disorder to help inform preemptive intervention approaches to avoid or prevent long-term posttraumatic stress and related disorders. This has led some researchers to study preventive strategies based on concepts of protective factors, including psychological hardiness and resilience. Their focus has been on risk and resilience factors (biological, cognitive-emotional, behavioral, social) that are considered important in the development and maintenance of post-trauma adjustment disorders in order to develop and test preventive interventions.

Conditions such as acute anxiety, depression, and PTSD can occur as a result of interacting individual and environmental circumstances. The nature, intensity and duration of exposure to trauma are clearly related to risk for adjustment disorders. Additionally, a host of other factors such as low socio-economic status, lack of education, previous trauma, adverse childhood and development factors, lack of social support, life stress, psychiatric history, family psychiatric history, posttraumatic psychiatric history, and peritraumatic emotion contribute to such risks. Sex and age are also specific risk factors involved in susceptibility.

Potential targets for selective prevention strategies include environmental interventions to limit or manage exposures, exposure-based interventions to increase familiarity with high probability events and cultivate accurate expectations about their impact, educational interventions to increase the controllability of acute stress reactions, shaping coping behaviors, and/or fostering help-seeking social interventions to build team/unit cohesion and bolster social support networks, and interventions to prepare family members for stressors introduced by their relative's deployment. A variety of intervention techniques or tactics for increasing high-risk individuals' resilience during short-term. intermediate, and long-term adjustment periods are needed.

Invisible Wounds

Since October 2001, approximately 1.64 million U.S. troops have been deployed for Operations Enduring Freedom and Iraqi Freedom (OEF/OIF) in Afghanistan and Iraq. Evidence has suggested that the psychological toll of these deployments—many involving prolonged exposure to combat-related stress over multiple rotations—may be disproportionately high compared with the physical injuries of combat. In the face of mounting public concern over post-deployment health care issues confronting OEF/OIF veterans, several task forces, independent review groups, and a Presidential Commission were convened to examine the care of the war wounded and make recommendations. Concerns have been centered on two combat-related injuries in particular: PTSD and TBI. With the increasing incidence of suicide and suicide attempts among returning veterans, concern about depression has also arisen.

The study discussed in a RAND (2008) report on PTSD, major depression, and TBI, not only because of high level policy interest, but also because, unlike the physical wounds of war, these conditions are often invisible to the eye, remaining invisible to other service members, family members, and society in general. All three conditions affect mood, thoughts, and behavior; yet these wounds often go unrecognized and unacknowledged, leaving a large gap in knowledge related to how extensive the problem is or how to address it.

Key Findings

- Approximately 18.5 percent of U.S. service members who have returned from Afghanistan and Iraq have PTSD or depression; and 19.5 percent report experiencing a traumatic brain injury during deployment.
- Roughly half of those who need treatment for these conditions, slightly more than half who receive treatment, get minimally adequate care.
- Improving access to high-quality care (i.e., treatment supported by scientific evidence) can be cost-effective and improve recovery rates.

Deployment Support Resources

A special collection of resources focuses on this topic. Several of these resources specifically concern deployment and how it affects military families. Others are targeted more toward helping children deal with war, terrorism, and a parent being away from home due to a deployment. Additional ones discuss suicidal behaviors.

Center for Military Health Policy:

- "Invisible Wounds: Mental Health and Cognitive Care Needs of America's Returning Veterans".
- http://www.rand.org/pubs/research_briefs/RB9336/index1.html
- http://www.cfs.purdue.edu/mfri/pages/military/deployment_support.html

Suicide as a Risk among Military Personnel

How can mental health professionals know which of the many variables in an individual's clinical presentation are most salient to that person's suicide risk? Such certainty requires an empirically validated prediction model that is specific to the population served. Data obtained through the U.S. Air Force (USAF) Office of Special Investigations and the USAF Institute for Environment, Safety, and Occupational Health Risk Analysis were analyzed using multivariate strategies of prediction based on an empirically validated model of suicide prediction (Brown, et al., 2000) and suicide completion versus non-completion status. The usefulness of the model to the USAF sample were discussed, and several factors unique to a military population were highlighted.

Implications

There are several implications of Brown et al's findings for clinicians. Given the increased stress and suicide risk factors in our society and the fact that suicide is considered to be one of the most common clinical emergencies faced by psychologists and other mental health practitioners, knowing what risk factors contribute most to a client's self-injurious behavior is essential to good risk management and client care. This information helps guide decision- making concerning hospitalization, referral, consultation, and the frequency of outpatient follow-up monitoring or intervention.

There are also potential implications for mental health professionals operating within industrial settings. For example, suicide risk assessments (and more general assessments of emotional stability) impact personnel selection, particularly for high-stress environments such as military, law enforcement, national security, etc. Unique to military settings, both individual risk factors as well as those that are systemic in nature are used to shape and support a healthier armed force. In keeping with this perspective, the USAF Surgeon General asserted that "we have to stop thinking of suicide prevention as something only mental health professionals do" (Redovian, 1999, p.1). This emphasis resulted in the continued development of surveillance tools (such as SESS) and prevention efforts (such as the USAF community awareness program). Through such identification and refinement of population-specific suicide risk factors, researchers and practitioners can create more focused prevention and intervention strategies. In most cases these strategies include a thorough biopsychosocial evaluation. Taking into consideration the variables outlined above can provide most practitioners with an excellent overview of potential risk factors in a client's risk profile. The ultimate goal of research on the subject of suicide is to one day be able to predict risk and avert clients from it. In the absence of such capability, we must continue to strive to reduce risk in the environment, identify those at risk through systemic prevention tools, and intervene in a way that minimizes the impact of the risk factors present.

Rozanov, Mokhovikov and Stiliha (2002) discussed the problem of suicidal behavior in the Ukraine military environment, and gave an example of the successful prevention approach they implemented. Their model of prevention is based on:

1. education of the responsible officers,
2. training of the representatives of the most vulnerable risk groups, and
3. follow-up procedures based on distribution of pocket books for soldiers, educational booklets, and sets of helpful material for officers.

One of their main conclusions was that the prevention activity must be organized as a continuum of actions, seminars, consultations, and materials distribution.

Chapter 8: What Is A Return To Equilibrium?

There comes a time when the military member (Regular, Reserve, or National Guard) returns from deployment and there comes a time when they seek to re-establish their lives. They attempt to return to what many refer to as "normality". However, it is very difficult to define what "normality" is or involves. However, a return to an equilibrium that incorporates their experiences is a more viable concept.

Balance

A mechanical system is at equilibrium if the forces acting on it are in balance. For example, when a body floats, the force of gravity is balanced by the buoyant force due to displacement of the liquid. The "balance of nature" (Pimm, 1991) is an extension of this idea to the natural world. The concept usually refers to steady flows of energy and materials, rather than to a system whose components do not change.

One analogy that can be used to demonstrate both balance and how it contributes to equilibrium is the balance scale that is often portrayed on or near our County Courthouse buildings. That is the balance scale held by Lady Justice. It represents not only how justice is balanced, but is also used to weigh many things in our world. To bring events and other things into a balance, we add or subtract from one or both sides of the balance scale. Normal or Normality is difficult to define and varies from individual to individual, from time to time, from culture to culture. When events disrupt the current normal, adjustments need to be made. These adjustments are made in order to bring the scale back into a state of equilibrium. As such, this equilibrium changes what we refer to as normal. Therefore, our normal has changed. Trying to re-capture what we referred to once as normal is no longer practical nor possible. Adding or subtracting to one or both sides of the balance scale represent adjustments to bring all into an equilibrium. Not making such adjustments can contribute to stressors and impaired adjustments to realities, resulting in difficulties re-integrating into civilian life following military experiences, especially those involving severe trauma and/or loss.

What Is Equilibrium?

In 1884, a French industrial chemist, Henri-Louis Le Châtelier (1850-1936), when commenting on chemical systems in equilibrium observed in 1884 that: "When a stress is applied to a system in equilibrium, the system will change so as to undo or offset the effect of the stress." He used this principle to describe the changes that occurred in chemical reactions at equilibrium when external changes were made to the concentration of the components of the reaction or to temperature or pressure, This principle has a much wider relevance and can be applied to a wide range of everyday situations, including any system in stable equilibrium. This also includes environmental systems in the political, cultural, scientific, technical, economic, and physiological areas. It can also provide a very useful analytical and predictive tool.

By applying Le Châtelier's Principle to systems in equilibrium, a system is said to be in *stable* equilibrium if, after a small perturbation is applied to it, it returns on its own to its original equilibrium state. An example of a system in stable equilibrium would be a ball at the bottom of a curved bowl. When the ball is deflected from the bottom of the bowl, other things being equal, gravity will always bring it back to the bottom of the bowl.

Not all systems in equilibrium are stable. An example of a system in *unstable* equilibrium would be a pencil balanced on its point. Any small deflection will cause the pencil to fall over. The fallen pencil will be in a very different equilibrium state from when it was balanced on its point and, therefore, the original equilibrium was unstable.

Le Châtelier's Principle did not require the system to return to its original equilibrium, nor is that required here. Application of Le Châtelier to situations outside of the chemical equilibria, to which it was originally applied, is only likely to prove to be valid where relatively small stresses and perturbations are being applied to the system in question.

A system in equilibrium is comprised of a number of inputs and outputs that have achieved a "balanced" state. In many every-day situations, the systems will be very complex and the inputs and outputs may be numerous. Because of this, a system under stress will not necessarily return to its original equilibrium state but may well return to a different one.

What Le Châtelier says is that the system will respond by resisting the changes and try to retain its original equilibrium state or at best achieve a slightly changed equilibrium state.

Most everyday situations we encounter are systems in some form of equilibrium. If they were not, we would not view them as a system at all since they would be changing continuously. We can use Le Châtelier to try to analyze and predict the outcome of other situations that affect us all. One such situation is actually occurring right now, where the equilibrium is already under significant stress. This involves the earth's atmosphere and global warming. This stress is being produced by our releasing into the atmosphere ever increasing amounts of carbon dioxide and other "waste gasses". These end up as a layer in the upper atmosphere and lead to the well-known "greenhouse effect", resulting in the average temperature of the earth's surface rising slowly. The impact of this if continued unabated will be catastrophic to the earth. Among the other bad things that could happen is that the ice caps would melt and large areas of the earth would become flooded. We would also get major changes in climate, etc.

A Corollary to Le Châtelier

A corollary to Le Châtelier's Principle called "The Goodwill Overshoot Principle", states that:

> "In the absence of sufficient damping force, when a system in equilibrium is put under stress, the offsetting changes that occur, following Le Châtelier's Principle, will likely overcompensate before equilibrium becomes re-established, i.e., the system will overshoot".

The equilibria we find in everyday systems are formed from fairly complex sets of inputs and outputs. Le Châtelier's Principle applies just as well to these situations as it did to the chemical equilibria for which it was originally prescribed. It should always be considered when evaluating the possible impact on the equilibrium of a system from stresses applied through changes to one or more variables, Doomsday scenarios are nearly always wrong because they do not take Le Châtelier into account.

As a direct consequence of Le Châtelier's Principle, it can be argued that it is usually safer to assume that a system that exhibits stable equilibrium is more likely to self-correct when small stresses are applied to it than it is to become unstable or to bring about fundamental changes to the system. Le Châtelier gave us this important piece of scientific observation and understanding, although in the limited area of chemical reactions.

The usefulness of Le Châtelier can be extended through the addition of the corollary of "The Goodwill Overshoot Principle" which states that in adapting to a stress, a system in equilibrium usually overcompensates in an attempt to sustain that equilibrium. As such, this can help in using Le Châtelier as a predictive tool when examining systems in equilibrium and under stress. This approach can be applied and used when analyzing and dealing with those who have been exposed to a traumatic event or critical incident that has disrupted their equilibrium.

PART V:
Wellbeing and Equilibrium

Chapter 9: Wellbeing

While psychologists have developed some models of behavior modification and mental reinforcement to help people achieve an elevated state of well being (Seligman, 2002) and economists have used theories of consumer behavior to give insights into the importance of relative income in explaining variations in well being (for example, Duesenberry, 1949; Friedman, 1957; Veblen, 1899; and Scitovsky, 1976), there have been fewer attempts to develop a more general theory of wellbeing that incorporates both economic and psychological factors.

Individuals achieve a higher state of wellbeing and happiness when they are in a homeostatic equilibrium. This equilibrium state has physical, emotional, psychological and environmental dimensions. Characteristics of this equilibrium include feelings of safety, trust, connectedness with friends, family and community, and a predictable and welcoming social and work environment. Individuals make decisions that help them move toward and achieve this state of equilibrium.

When individuals are displaced from this equilibrium as a result of abrupt and strong changes (shocks, trauma) in the overall environment, wellbeing and happiness are affected. If the shock is positive (e.g., marriage, birth of a child, promotion, windfall inheritance), the individual experiences an increase in wellbeing. If the shock is negative or traumatic (e.g., death of a child, parent or spouse, falling seriously ill, demotion or getting fired), wellbeing is adversely affected.

Behavior adjusts to restore the individual to homeostatic equilibrium through a combination of physical, emotional, behavioral and psychological adjustments. This behavioral readjustment is known as *allostasis*. This process of adjustment to external shocks/traumas creates physical, emotional and psychological stress.

The concept of homeostasis as it relates to human, ecological and physical systems pays particular attention to the impact of external shocks/traumas (e.g., unemployment, deteriorating health, divorce, emotional challenges) on agents and how agents react to these shocks or traumas. Behavioral adjustments by these agents are then related to the concept of homeostasis and allostasis, and to research findings on the determinants of happiness and wellbeing.

An aspect of the relationship between risks and the homeostatic equilibrium relates to behavior that has been referred to as risk homeostasis. This behavior implies that there is an optimum or equilibrium level of risk that people are generally comfortable with. If this is true, then efforts to decrease risk may be met by riskier behavior. Consider the case of farm tractors and road design. When tractors were designed for greater stability, farmers used them on steeper slopes and the accident rate remained constant. When highways were designed to be safer, drivers increased their speed and took more risks and the accident rate remained at previous levels (Slovic, 1984). Ample evidence also illustrates that when individuals are placed in dangerous situations where there is risk of injury or death, they experience a higher level of stress (Seligman, 2004; Spitzer et al., 1995) which implies that they have the urge and motivation to take action to return to their equilibrium level of stress which is consistent with their desired comfort zone.

The nature of an individual's disposition also affects the rate of return to the equilibrium set point. For example, research suggests that optimistic patients live longer than pessimistic patients (Palmore 1969a, 1969b). Happy people recover faster and some diseases can be cured or treated more effectively when the patient has a happy and upbeat attitude (Diener and Seligman 2004). As a corollary to this, a pleasant mood seems to lower blood pressure and that a high level of stress reduces the ability of the immune system to fight off disease. Furthermore, depression and anxiety, two major forms of mental illness, lead to significant declines in well-being (Spitzer et al 1995; Packer, Husted, Cohen and Tomlinson 1997). On the other hand there is evidence that happy people show low signs of mental illness (Diener and Seligman 2002). In addition, duration of unemployment and wellbeing are negatively related, and absenteeism and turnover rates are lower when workers are happier (Clegg 1983, Clark 2001; Akerlof et al. 1988).

There is also ample evidence that agents who experience negative shocks or traumas that reduce wellbeing make persistent attempts to return to the homeostatic equilibrium and to increase their levels of wellbeing and happiness. Those who are unemployed look for work. Those who are sick go to the doctor. Those who are divorced begin to date and often remarry. Those who move to a new city or neighborhood make efforts to make new friends.

While the desire to return to a set point or homeostatic equilibrium may not always by synonymous with an increase in happiness and wellbeing, the various behaviors described above do suggest that the motivation to return to such an equilibrium set point is strong and highly desirable in a wide variety of circumstances, including war settings, disasters and other critical incidents. This behavioral pattern generally reflects a desire to return to or move toward the familiar, predictable and comfortable, and away from undue stress and aggravation. To the extent that these states of equipoise and relaxation are preferred, we can infer that they are also positively associated with higher levels of wellbeing and happiness. When life circumstances such as stress, trauma, critical incidents, war, death of relatives/close friends and other negative social developments disrupt this equilibrium, wellbeing is compromised.

Some researchers have pinpointed the importance of such a homeostatic set point in determining the level of wellbeing. For example, Cummins and Nistico (2000) suggest that life satisfaction responses are not free to vary over the full range of possible outcome (e.g., from being very unhappy to very happy). The distribution of responses tends to be confined to a narrow range. They argue that this is because of the operation of such a homeostatic mechanism. They suggest that high self-esteem, control, optimism about life's circumstances and the understanding that we are in control of our own destiny help to constrain responses to a narrow range. This optimistic approach to life in general serves as a psychological buffer against misfortune, if and when it arises. Furthermore, this narrow range of variation in average subjective wellbeing responds quickly following misfortune. Even those who have suffered serious accidents involving paralysis return to a fairly optimistic view of life after some time. Such a view of behavior stresses the powerful psychological forces that lead to a return to a set point or homeostatic equilibrium. Psychologists refer to this belief pattern as positive cognition bias. This bias leads to a variety of observed behaviors that buffer the psyche. These include people's need to preserve their self esteem by downplaying the ability of others; to control essentially random outcomes by mental concentration or to predict many more positive than negative outcomes when asked to think about the future.

According to Pimm (1991) and others, long return times may be indicative of a loss of resilience. Pimm is concerned with behavior near a stable equilibrium. In that case, a long return time for a given displacement from the equilibrium indicates a small coefficient or, equivalently, a small derivative. Others are concerned with behavior of a system with two or three equilibria, one of which is stable. Resilience describes the tendency of the system to return to its stable equilibrium. A long return time is due to disturbances that bring the system near an unstable equilibrium, or possibly to a weak repulsion from an unstable equilibrium.

One should not interpret the strong desire to return to a set point as a reason to forsake explanations for variations in observed wellbeing. Models which try to describe the causes of wellbeing are only successful in accounting for a small proportion of the variation in individual responses. Analyses suggest that happiness over time varies directly and significantly with several dimensions of people's lives including family life, health, work and social environment (Easterlin 2004).

Life disruptions have a strong negative and very strong impact on happiness and wellbeing. Social mobility as reflected by the number of moves has a significant impact on wellbeing. Making more moves tends to reduce trust, while fewer moves tend to increase trust and social cohesion in neighborhoods, towns, etc. Twenge (2002) and Magdol (2002) concluded that frequent moves have a detrimental impact on wellbeing. Illness, mental anguish and death in families also have a very strong negative impact on wellbeing (Di Tella et al 2003), Layard 2005; Diener and Seligman 2004). Disruption in the form of vulnerabilities to floods, hurricanes, earthquakes, tornadoes, and other natural disasters, also results in lower levels of wellbeing (Veenhoven 1994). Poor health and illness diminishes wellbeing quite dramatically as shown in several studies (World Values Study Group, 1994; Packer, Husted, Cohen and Tomlinson, 1997; Diener and Seligman 2004; Gerlach and Stephan 1996). Depression and other psychiatric illness together make up about 30 percent of the various causes of disability. This is a much higher rate than disability from alcohol and drug addiction (together 10%), respiratory illness, cancer and heart trouble (together 15%) (Layard 2005, Cahpter 11). There is also evidence that happy people show fewer signs of mental illness (Diener and Seligman 2004).

Rewarding social interactions are key components of wellbeing (Baumeister and Leary, 1995) This involves frequent and pleasant interactions with a few people within the context of a stable, trusting and mutual caring environment. Ongoing relationships, within a framework of mutual concern, provide a stronger and more substantive bond and feeling of belonging than one based on self interest alone (Clark 1984; Clark and Mills 1979). Superficial social contacts cannot substitute for deeper and more intimate relationships (Weiss 1973; Baumeister and Leary 1995). Positive social bonds are associated with positive emotions and higher levels of wellbeing (see Sternberg 1986 and McAdams 1985). Conversely, the loss of friends leads to loneliness and depression (Leary 1990) as well as anxiety (Baumeister and Tice 1990). Other research shows that intimate relationships and close social and family ties are highly valued by respondents and, in the case of sexual intimacy, results in a significantly high increase in wellbeing. (Kahneman, et al., 2004; Diener and Seligman 2002; Blanchflower and Oswald 2003).

Two specific events that have a strong impact on a person's need to belong are divorce and death. Even though marriages that end up in divorce court may not have been joyful, divorce nevertheless results in negative feelings and reduced wellbeing (Weiss 1979; Price and McKenry 1988). The death of a spouse, child or close friend rank high on the list of stressful and difficult events and can result in a period of depression (Holmes and Rahe 1967 and Weiss 1979).

People are adversely affected by negative disruptions such as illness, unemployment and divorce yet they recover more rapidly than they expect from these disruptions. People value their freedom and are adversely affected by rigid constraints and governmental controls. They value friends, family and social relationships more than work and commuting. Extra income increases happiness less as people get richer. Motivation for maintaining the status quo (being comfortable) is consistent with homeostasis and is reinforced by social nature and the search for trust.

When trauma occurs, people are thrown so far out of the range of equilibrium that it is difficult for them to restore a sense of balance in life. Trauma may be precipitated by stress: "acute" or "chronic". Acute stress is usually caused by a sudden, arbitrary, often random event. Chronic stress is one that occurs over and over again—each time pushing the individual toward the edge of his or her state of equilibrium, or beyond. Most trauma comes from acute, unexpected stressors such as violent crime, natural disasters, accidents or acts of war. Some trauma is caused by quite predictable (but hated) stressors, such as the chronic abuse of a child, spouse, or elder abuse. "Developmental crises" come from transitions in life, such as adolescence, marriage, parenthood and retirement.

Before a disaster occurs, the majority of people are usually in a state of relative equilibrium. When a stressful incident occurs, people experience a state of disequilibrium and there is a perceived need to return to equilibrium. Faced with this dynamic situation, the role of mental health professionals and related responders is to support, and at the same time, accelerate the re-equilibrium process. Thus the worker may have to deal with disaster victims who are in various state of recovery. On the one hand, there are those who, because of the presence of compensation factors, situational support, and appropriate adaptation mechanisms, have a realistic perception of the incident. This usually leads to resolution as one's equilibrium is regained without crisis. On the other hand, there are others who, because of the absence of compensation factors, of situational support, and of adaptive mechanisms, may have a distorted perception of the incident. Problems then remain unresolved resulting in disequilibrium and crisis. A crisis intervention referral is needed (Aguilera and Messick 1976).

When a disaster or other critical incident happens, people are traumatized at different levels and much discussion revolves around returning to normal. However, "normal" has changed. It is a concept with many different meanings. It tends to be a term that it is very difficult to define and, consequently implement. What was "normal" before can never be the same again. What can happen and what is constructive, and more easily defined is "Return to Equilibrium". This involves integrating the event, its effects and its meanings into one's life and recognizing it as now part of one's life. Taking that and building a new balance in life can bring one into a new and enriched life. To use a concrete example, it is quite akin to taking an old-fashioned balance and adding or subtracting to one side or the other in order to gain a balanced scale. Experience in life changes each of us whether the experience is good or bad. Integrating that experience into our life creates that new balance. It is a changed normality that results from the new balance and a return to equilibrium.

Returning to Equilibrium following a major hurricane or other natural event is different from Returning to Equilibrium following a man-made traumatic event, critical incident, terrorism event or war. However, in all cases, re-establishing a balance in life that integrates the event as part of one's life and moving into the future in a constructive manner is what develops our new "normality". Such a "Return to Equilibrium" is a goal of recovery.

A stress response that fails to return to a state of equilibrium becomes unresolved psychological/emotional trauma. Emotional or psychological trauma is the extreme end of the stress disorder continuum. It is stress run amuck—a deregulation of the nervous system that remains fixed and contributes to lifelong mental, emotional and physical disorders including anxiety and depression. Emotional or psychological trauma can result from such common occurrences as an auto accident, the breakup of a significant relationship, a humiliating or deeply disappointing experience, the discovery of a life-threatening illness or disabling condition, or other similar situations. Traumatizing events can take a serious emotional toll on those involved, even if the event did not cause physical damage.

Following a disaster or critical incident, it is common for individuals to feel pressure to reduce distressing emotions brought on by the event and return to adaptive, independent everyday functioning. For responders, returning to everyday functioning involves managing residual effects from the event, being able to handle the day-to-day stressors of work, and being prepared to respond to any future crisis. Responders may want to avoid talking about the event. Avoidance can also take on other forms, such as missing work or even working too much. Although avoidant strategies may be initially highly adaptive, allowing the responder to continue working, continuous use or overuse of avoidance can impede long-term positive readjustment. An active problem-solving approach that actively addresses difficulties posed by the stressor can help with both chronic and acute stressors. For example, finding meaning in some outcome of the event may minimize feelings of helplessness, instill a sense of control and mastery, and is typically associated with better physical and psychological outcomes than avoidant coping.

Following an intensely stressful event, responders may believe that by talking about the event they will burden family and friends or that other people "just don't understand". However, talking about, or even writing about, the experience can be healthy and therapeutic. Discussing the event does not need to take place immediately following the stressful experience. Judgment should guide the decision about to whom and when to open up.

If a trusted confidante extends an offer to listen and discuss the experience and reactions to the event, and the responder desires to talk about his or her experiences, this might be beneficial. If, on the other hand, responders are not yet ready to discuss their experiences, their decision should be respected. It is important that all relationships be built upon communications, including the right to not communicate. It is helpful, however, for responders to remember to keep their options open and to let those who have offered their support at least know that their offer has been heard and appreciated. This is especially true with family members.

Every time a stressful event happens, there are certain recognized compensating factors which can help promote a return to equilibrium. These include:

- perception of the event by the individual
- the situational reports which are available
- mechanisms of adaptation

The presence or absence of such factors will make all the differences in one's return to a state of equilibrium. The strength or weakness of one or more of these factors may be directly related to the initiation or resolution of a crisis.

When stress originates externally, internal changes occur. This is why certain events can cause a strong emotional reaction in one person and leave another indifferent.

People in crisis are extremely vulnerable. They are open to hurt as much as to help. The goal of crisis counseling should be to protect them from further harm, while providing them with immediate assistance in managing themselves and the situation. Counselors can provide brief, clear, and responsive directions and support for those affected by traumatic incidents.

Provide Information Sessions

Information sessions presented jointly with the organizations involved (e.g., military family support group groups) are intended for the whole community. They consist of providing general information and dealing with services available, and problems associated with reintegrating into and returning to civilian life. During the information sessions, the following messages are among those given with regard to physical and emotional reactions:

- the physical and emotional symptoms are part of a stress

reaction and are considered normal;

- these symptoms occur in most people in a situation of stress, threat or loss. They are primitive reactions of the mind and body, and their purpose is to help the individual survive;
- stress syndromes, although normal, can, however, present health risks if they persist, since they rob people of energy and make them vulnerable to illness, In some cases, they can even have repercussions on a person's whole life;
- there are many ways of dealing with stress reactions, such as surrounding oneself with people one feels good with and with whom it is easy to talk about what one is experiencing, doing vigorous physical exercise, or using relaxation techniques.
- the most effective way of relieving stress reaction syndrome is verbalization sessions on the event.

Mental Health Services: Women in the Military

Katz et al (2007) studied eighteen women who served in Operation Iraqi Freedom/Operation Enduring Freedom (OIF/OEF) and sought mental health services at a Veterans' Affairs (VA) medical center. Ten of the 18 women (56%) reported military sexual trauma (MST) while serving in OIF/OEF. All 10 with MST reported sexual harassment, 6 of the 10 (33% of the sample) reported unwanted physical advances, and 3 (17%) reported completed assault or rape. Fifteen women also completed a questionnaire about their experiences and the Iraq Readjustment Inventory (IRI) developed for this study. High reliability and high correlations with clinician ratings make the IRI a promising measure for future research. A comparison between those with and without MST revealed that those with MST had higher clinician ratings and IRI scores, suggesting greater difficulty with readjustment. While MST was significantly correlated with clinician ratings and readjustment scores, the variables "being injured" and "witnessing others injured or killed" were not. These preliminary data suggest that MST OIF/OEF women seeking mental health services is a critical factor for predicting symptoms and difficulty with readjustment to civilian life.

Counselor Preparation

Exploratory research by Pearlman and MacIan (1995) showed that trauma counselors who were newer to the counseling field and those with a past history of trauma exhibited symptomatology similar to individuals with PTSD. This is identified as secondary trauma stress or vicarious trauma. Pinto (2003) explored how conducting counseling with clients who have a history of trauma impacts the generalist counselor. The first goal was to explore how a counselor's level of experience and personal history of trauma influenced their level of secondary trauma. A secondary goal was to identify the differences that level of experience had upon their manner of coping. Eighty-two participants (49 novice and 33 expert counselors) were recruited from Southern California. 75.5% of the sample was Caucasian with 10.2% Asian-American, 6.1% Latino, 2.0% Pacific Islander, and 6.1% unidentified. The sample consisted of 34 male (34%) and 58 female (71%) graduate students and/or counselors currently participating in counseling related activities. Participants indicated their experiences with primary and secondary trauma and completed a survey packet. The packet contained a series of questionnaires assessing their level of secondary traumatic stress, posttraumatic stress, burnout, and means of coping. These include:

- Figley's (1995) Compassion Fatigue And Satisfaction self-test,
- Weiss' (1996) Impact Of Events Scale-Revised,
- Pearlman's Belief scale, and
- Folkman and Lazarus' (1988) ways of coping questionnaire

As hypothesized, the results suggested that one's history of trauma influences one's level of secondary traumatic stress. No differences were found for burnout, vicarious stress, and posttraumatic stress when analyzed based on level of experience. These results suggested that differences may exist among the concepts of burnout, posttraumatic stress, and secondary traumatic stress. This lends additional support to reconceptualize the category of PTSD to include PTSD and secondary traumatic stress. Further, the results suggested that novice counselors use distance as a coping strategy and those with a primary and secondary history use confrontive coping and positive reappraisal coping strategies to deal with clients with a history of trauma.

Some Suggestions for Coping

Verbalization sessions are a simple but effective method for providing assistance in coping with and readjusting to civilian life. A verbalization session helps alleviate acute stress reactions in order to reduce or prevent delayed stress reactions. This method is a rational way of dealing with stress reactions. The intervention model focuses on 3 specific objectives:

- to help those affected express their feelings;
- to assist them in understanding their emotional reactions and their behavior;
- to promote a return to a state of equilibrium in each individual.

Specialized literature in this area suggests that this type of intervention generally provides good results. The optimal intervention should ideally take place as soon as possible after the appearance of symptoms (the concept of immediacy) and as close as possible to the site of the incident (concept of proximity). This may not be possible in the case of many in the military. However, it can take place within a military setting upon return. It should bring together similar groups (concept of community) and create a climate that carries a clear message: what they are experiencing is normal; it can be healed and they will be able to adjust to their new level of equilibrium (concept of expectancy).

Groups should be homogenous, the atmosphere should be positive, supportive and understanding. Verbalization sessions should be led by competent mental health professionals who are knowledgeable about such interventions and who have received the necessary training.

Follow-up

Organizations should have some means of monitoring individuals' recovery. This can take place as a routine series of follow-up meetings. During re-integration to civilian settings, this can include meetings between employers and employees, or a routine medical check with a health nurse or physician or mental health professional. The purpose of the follow-up is to allow the returning veteran further opportunity to talk about feelings and readjustment and re-integration into civilian life. It is also to assess with the individual whether the symptoms are diminishing. A good time to do routine follow-up is within a month to six weeks after return home. If veterans still have difficulties with stress symptoms at that time, a routine referral to a mental health counselor should be suggested. A pre-established plan for referrals to counselors who are knowledgeable or specialize in working with returned veterans and their families should be developed. Some suggested plans that deal with this include TriCare, Veterans Administration and Veterans Groups.

Chapter 10: Where Do We Go From Here?

Traumatic Brain Injury (TBI)

One area that we have presented limited information about in this volume is Traumatic Brain Injury (TBI). TBI is defined as damage to the brain resulting from external mechanical force, such as rapid acceleration or deceleration, blast waves, or penetration by a projectile. Brain function is temporarily or permanently impaired and structural damage may or may not be detectable with current technology. TBI can be caused by falls, vehicle accidents, and violence. It can cause a host of physical, cognitive, social, emotional and behavioral effects and outcomes can range from complete recovery to permanent disability or death.

Some of our deployed service members have sustained attacks from explosions or blasts by rocket-propelled grenades, improvised explosive devices (IEDs) and land mines. Since October 24, 2008, the Veterans Administration has increased the disability rating for TBI Veterans. Depending on the extent of the injury, Veterans are eligible for up to 100% disability rating. TBI is the signature injury of the wars in Iraq and Afghanistan. One promising technology for treatment is called activation database guided EEG biofeedback. It has been documented to help return a TBIs auditory memory ability to above that of the performance of control groups. There have been a large number of studies and reports that deal with various forms of TBI. In order to effectively review and present the information included in these will entail an additional volume that is planned for the near future.

Trained Medical, Nursing, and Psychological Personnel

Another area of concern for our returning Veterans, their families, and for our country in general, involves trained personnel. The military has trained nurses, Medics, doctors, and other medical and psychological professionals. These Veterans not only have training in their professions, but also are experienced. It is difficult for many of them upon return home to get the needed licenses and certifications to be able to use their skills and trainings in civilian applications. They are required to undergo trainings and education which they already have through military and their military experience. They are required to redo such trainings in many circumstances, costing time as well as money.

There are no current ways that they seem to be able to get credit for the trainings and experiences they already have by either College Level Examination Program (CLEP) testing or other means. As they already have most of the trainings as well as experience, they need to be able to get credit for what they have already accomplished. In the western states in rural and frontier areas (e.g., Wyoming, Nevada, Arizona, New Mexico, Montana, Idaho, Alaska, and other areas), we have a great need for both medical, nursing, psychological, and other trained professionals to provide services for not just our returning military, but also for civilians in those areas. Many of our small and isolated towns, communities, clinics, etc need such professionals. Too often people in these areas may need to travel 100 or more miles in order to get professional help.

A system that includes already trained, experienced professionals can help fill those gaps. A system of traveling professionals and/or traveling clinics with such professionals would be able to provide such services cost-effectively to those who need and seek such help. Such a system can be effectively backed up by computer and communication systems with specialists and other assistance from larger communities with larger hospitals when needed. The personal contact by medical and psychological professionals is very important and much appreciated in rural communities. When the professional presence comes to and knows the community, the problems the community deals with and the people living there, the health care needs and those providing them are really appreciated. They are part of the community. They are not just someone who may see someone for a 15 minute block of time after traveling many miles and often involving overnight stays at a motel. They are not dealing with someone in a "medical ivory tower". The community provider is someone they know and talk with who is someone who understands the community, not just an entity to whom a large fee is paid who has very little knowledge of the rural concerns. They most likely know their profession well, but they often have no time to become part of the small rural/frontier community. Rural community practitioners can find ways to consult with those who live and work in the larger settings. One asset for these rural/frontiers are our returning trained military Veterans. This is a resource and opportunity for all that should not be overlooked or taken lightly.

Additional Areas of Abilities and Educational Possibilities

Other areas that many of our Veterans have trainings in include engineering similar to the work done by the Army Corps of Engineers. With disasters resulting in flooding (e.g., Katrina follow-up, Hurricane Sandy and nor'easter damage done to the east coast and other areas) we have an opportunity for these men and women to help repair, build, and re-build the infrastructure in many of these areas. They have the skills, the abilities, the training to do so. Others can increase these skills through education as well. We need people who know how to build roads, even railroads, other transportation systems to improve those we have and to build the future of this country.

Our Air Force and Naval Aviators know how to fly the aircraft, not just passenger, but also cargo. They know how to build, man, and maintain newer and better airfields and terminals. They know how to do so in the face of all kinds of adversity and weather.

Our rural/frontier areas need Veterans who can improve our agriculture in new ways to enhance continued and ongoing growth and development of Agriculture that includes crops, livestock and improved distribution of foods and other agricultural products.

Planning For the Future

All of the above are just a start. With our returning Veterans and others in the civilian sector, we are poised to build for the future by adequate long-range planning. We need to plan for what we want for the future, not just take for the present and forget the future. We can do it and our Veterans and others have the chance. What is the future that we want and plan for? They are all part of the *Next Great Generation*. We are a great country. We can make it even better for all.

Chapter 11: Development, Implementation and Use of Teleconferencing in Rural/Frontier Settings

"Few things can help an individual more than to place responsibility upon him and to let him know that you trust him."
— Booker T. Washington

Teleconferencing in its various forms helps provide different methods for improving communication and enhancing interactions between the professional and clients. If used appropriately, these techniques can not only improve interactions in a timely manner, but also reduce costs involved in the provision of such services for clients as well as the professionals. Familiarity with and comfort in the use of such techniques and technologies can help facilitate these interactions. Following are a number of areas to take into consideration when designing such a center.

Video and Audio

When video is used in conjunction with telephone or other types of high-quality audio on channels for conferencing or therapy-related purposes, the technology has proven to be useful (Baer, Cukor, and Coyle, 1997). Compared to in-person appointments, teleconferenced interviews were found to have satisfactory levels of client satisfaction and clinical outcome (Rohland, 2001). Dongier et al., 1986) found that most patients rated videoconferencing sessions as above-average. The consultants, however, disagreed. They rated global assessment, diagnosis, and written consultation by videoconferencing as being slightly inferior.

Cost Effectiveness

Important factors here include decisions involving telephone and/or video-conferencing. For example, reducing a client's travel expenditures doesn't affect reimbursement of the clinician or directly benefit a third-party payer. Prior to investment in special equipment and communication services required for videoconferencing, the clinic should weigh the financial advantages in terms of the benefits to the client. In most places, remote mental health treatment tends to be confined to videoconferencing done under certain defined circumstances. Use of electronic communication is best applied when it serves the best interests of the client.

The cost savings of using telephones for psychotherapy were established in the 1980s and 1990s. This may be changing as new models of phone technology develop that combine telephonic with web-based services and information. Feasibility studies have demonstrated renewed interest in telephone counseling. Many of the newer cell phone types such as Blackberry and Android are a few examples. Research and the market can provide useful information about the cost-effectiveness of using such devices in the delivery of mental and behavioral health services. Factors to consider in adopting such a system include ease of use, cost-effectiveness and quality of care. In a community-based study, utilizing video-conferencing was found to be economically advantageous as compared to in-person consultation for clients and patients remote from such services. A cost saving was reported to average $83 for one-way travel (Brown-Connolly, 2002a). Researcher reported benefits of using teleconferencing included time involved in consultation and reduced need for travel for nursing home residents)Johnston and Jones, 2001). In another area, the National Institute on Disability and Rehabilitation Research's Videocounseling for Rural Teens with Seizure's Project has sought to assess the cost-effectiveness, impact on family relationships, and improvement of specific problems for home-based video, speakerphones, and office-based counseling (Center for Research on Telehealth and Healthcare Communications, 2002).

Costs for full video equipment continues to drop. In some cases, sharing facilities and/or equipment can help with acquisition and maintenance. community colleges and community centers are some examples of such resources. Following are some factors that should be considered:

- Telephonic Practice Enhancers: Recent enhancements in telephone technology have provided many additional means of correspondence and communication. For example, most phones are capable of sending and receiving text messages. Telephone texting can be very effective in promoting client compliance (Neville, Greene, Mcleod, and Surie, 2002, p. 600).

- Voice Over Internet Protocol (VOIP): Skype and similar software.
- **Videoconferencing Practice Enhancers**
  - Transmission Channels—Distinguish between telecommunication transmission channels and telecommunication devices.
  - Bandwidth limitations in low-speed transmission channels make it difficult to transmit audio and visual signals at a quality that adequately approximates in-person communications (Fussell and Benimoff, 1995).
- Video Quality: A good-quality image may boost the effectiveness of psychotechnology beyond what may be ordinarily achieved with standard in-person encounters. The quality of the video-conferencing will remain currently dependent on its treatment uses.
- Broadband Connections: Broadband Internet connections at home, in school and at work are providing more available opportunities for practical general mental health care using low cost off-the-shelf equipment. An example of a simple system would be Skype or a related system with security enhancements to provide for confidentiality. A similar general system could be very effective in certain circumstances for many of our returning Veterans in rural and other remote areas.
- Videoconferencing Devices and Programs: Quality of video-conferencing technology depends on resolution, color depth, frame size, and frame rate. Frame size or making a picture larger doesn't necessarily make it clearer or more detailed (Whatis.com, 2001a).
- Videophones: Videophones are small, easily installed units that attach to a telephone and some type of monitor (computer or television). The telephone unit includes a camera that allows a remote party to see the message sender in real time. Many phone devices and newer computers (e.g., laptops) contain cameras as part of their system.
- Practicalities—Successfully using videoconferencing technology in mental and behavioral health settings is dependent on a number of

factors. These include telepresence, technical audio and video considerations, and cleaving to videoconferencing etiquette (Barker and Alessi, 2001). These are also limited by the kind of videoconferencing software and the equipment available.

- Telepresence—Mental health professionals must be aware of the remote site as being part of their clinical milieu. For example:
  - Who is in the room? Only persons known to all parties and announced should be in the room. If other people enter during a consultation, they need to announce their presence.
  - Does the client have privacy during a therapy session? Unwittingly, clients may compromise their own privacy. For example, an overly enthusiastic client at work in an office cubicle may be overheard by coworkers.
  - Are the rooms in both locations sufficiently soundproof?
  - Is on-site support staff available?
  - Backup services for client support should be identified before routine remote consultation is started.
  - Emergency support services should be available for protection. (Anderson et al., 1996; Cukor et al., 1998; Prussog, Muhlbach & Bocker, 1994; Sellen, 1995).
- Referrals, Client Education, Consent—Methods and approaches for safeguarding and delivering each of these needs to be developed and tested out before implementing.
- Delivery of Care—Develop techniques and methods for this. What about translations for those in additional languages.
- Conferences, Workshops, Consultation—These represent some very cost-effective methods for training for both presenters and attendees. A great way to provide Continuing Education.

Ethics and Practice Management

"First of all, do no harm."

—Hippocrates

Ethics

Harking back to a dictum uttered centuries ago in Ancient Greece, our human service professionals seek to emulate the high standards set by their forebears. Our modern civilizations have mostly sought to provide laws that govern their populations. However, the ethical codes of the various professions are generally set higher than the population in which they live. This is at least partially due to the fact that such professionals have a greater impact on the quality and sometimes length of the lives that they serve. Ethics involves the application of human moral principles to certain codes of behavior that are usually adopted formally by associations or organizations (Ladd, 1991). It is the responsibility of each professional to decide appropriately how to deal with various ethical dilemmas when using the ethical code specific to the relevant professional association or organization (Pope and Vasquez, 1998).

While ethical codes are a level above the law, their written rules are legalistic and often serve as a basis for, or influence, legal decisions as well as public policy. For example, the APA "Statement on Services by Telephone, Teleconferencing, and the Internet" (American Psychological Association, 1997) is one type of document that could influence public policy.

The uses and adaptations of teleconferencing and other forms of *telehealth* technology are not likely to change the standards of professional conduct expected among professionals utilizing such innovations. As behavioral health care establishes outposts in cyberspace, some innovative concepts will undoubtedly stretch professional ethics. Mental health professionals should not disparage the adventurous clinician who explores new online approaches. Despite the unique aspects of clinical care, some groups have been developing guidelines for professional conduct in new practice areas related to service delivery through telehealth (Milholland & Reed, 1998). One example of a standard of professional conduct that is common to all health care professionals concerns their responsibility to protect client confidentiality. The use of telehealth technology, such as Internet-based data transmission and electronic medical and psychological records, may create serious difficulties with professionals' ability to meet standards of professional conduct in this area, and need to be examined in this light.

One example of such a dilemma can occur during response phases of mental or behavioral health during or following a traumatic event. Communication in such settings is often either unavailable or extremely erratic early on. Cell phones, computers and other common forms of communication are often first to be adversely affected. This can result in the use of two-way hand-held radios as a major method of communication between and among a base or headquarters and mental health and family service workers in the field. A concern voiced by such professionals following a past major destructive hurricane involved the confidentiality of the person they were talking with in the mountains. Of course, confidentiality is critical. However, when there are lives involved, a command decision needs to be made concerning priorities. In western Puerto Rico, following Hurricane George, phone or other forms of electronic communication were non-existent during the first few weeks, especially in the mountain communities. Mental health personnel were concerned about seeking consultation via hand-held radio communication. They were concerned about confidentiality. The mental health workers needed advice and feedback concerning mental and behavioral health issues. They were many miles from headquarters and unable to transport victims of the hurricane. There was an immediate need to discuss issues. A command decision was made to discuss via radio. No other form of communication was available or functioning. The situation was a crisis in a disaster setting. The likelihood of a break in confidentiality was minimal to non-existent. Advice and consultation took precedence.

Mental health professionals who are thinking about using teleconferencing and tele-mental health technologies should review current ethical guidelines when considering practice innovations. The technological landscape is so diverse and changeable that most codes and rules are subjected to complaints, appeals, and desperate searches for loopholes by professionals and clients alike. Mental health professionals have only begun to collect data about the degree to which members believe in or comply with their professional associations' standards of conduct (e.g., Pope, Tabachnick, & Keith-Spiegel, 1987). Data are not readily available to inform either the clinical decisions of individual practitioners or the attempts of relevant professional associations to extend formal standards of practice into new areas, such as online clinical practice.

Clinical standards cover professional conduct, practice and clinical treatment guidelines, standards of care, scope of practice, and other related issues. Standards and guidelines are rules promulgated by professional organizations and associations and by professional licensing boards, which in many cases are government agencies. In the United States, state licensing boards enforce legal and regulatory requirements and regulate areas of professional practice. Additionally, professional associations police their own ranks. The standards of professional conduct that a mental health care organization sets forth define the responsibilities for which the organization holds its members accountable. They may take action against members who do not adhere to their ethical standards by means of sanctions, expulsion, or reporting them to the appropriate state board. Due to ongoing skepticism concerning mental health delivery over the Internet, practitioners must be aware of the evolving laws. Additional concerns of importance include malpractice coverage and liability (Office of Technology Assessment, 1990), the ramifications of practicing outside what some practitioners deem the current "standard of care", emergency backup of data files, storage of videotape of therapy sessions, and proper documentation of services delivered (Brandt, 1996). Another area includes precautions concerning treatment of minors without parental consent and regulations concerning patient/client rights (California Healthline, 1999.

Ethical Principles for Online Clinical Practice

In response to unprecedented growth of the Internet in the past decade and a half, some groups have set forth ethical principles for health care delivered through the Internet (Wooton & Blignault, 2003). These principles cover advertising, commerce, partnerships, and practice for both general and specific types of health care delivery (Eng, 2001). Without established rules governing the rapidly evolving area of distance therapy in a particular mental or behavioral health discipline, practitioners must rely on basic ethical precepts. Once formulated, online standards should undergo continued review and revision (Plaut, 1997). Similar to in-person sessions, videoconferencing may need to be recorded. Recording may create unnecessary interpersonal, procedural, and legal complications. HIPAA and other federal and state laws will govern the use, disclosure, and storage of video conferenced sessions. Video recording may fall under the definition of "psychotherapy notes", which are given special protection under HIPAA. Such considerations need to be outlined ahead of time and spelled out for clients.

While professionals are bound by ethical principles, certain extenuating circumstances presented by psychotechnologies may obstruct compliance with these principles. For example, most psychotherapists must explain the limits of confidentiality at the onset of service provision. This initial step may be complicated for a therapist who has a poor understanding of the technology and how it affects the security of information. In a similar vein, the safeguarding of professional records may become difficult when a therapy session is videotaped by an employer who owns and creates backup data files for video equipment the therapist uses.

The process of service delivery can shift dramatically in various situations. An example would include the provision of emergency interventions delivered remotely. There is a current dearth of information about and standards for providing emergency remote mental health services. An emergency may develop during the course of ongoing distance therapy. Disaster situations are one good example (Puerto Rico and Hurricane George follow-up and other critical incident responses are examples).

Confidentiality

Whatever the situation, whatever the technology, practitioners should always apply common sense. By making massive amounts of information readily available and easy to store, retrieve, and search, the psychotechnologies have highlighted the importance of many confidentiality issues. For example, must video via Internet be encrypted or otherwise secured to meet HIPAA standards? (yes). Are chat rooms confidential? (Not typically, despite website owner claims.). Many other ethical issues are involved in the collection, storage, transmission (electronic or physical), and use of confidential data and their incorporation into records. Even if practitioners meet all basic requirements, some surprising dilemmas can surface. For example, how much access should be given, and exactly what powers should be entrusted to law enforcement officials (Bayer & Colgrove, 2002)? Should the confidential mental health files of a parent violating a court order for a child support be used by an enforcement agency to locate the delinquent parent (Ziglin, 1995)? When should a health professional balk at a demand for information and turn to the courts to overturn the demand?

The following are a few of the factors practitioners should be aware of and probably use as major guidelines:

- **Avoiding Harm**—Serious potential for damage to online clients can arise from many sources

- **Managing Crises and Emergencies**—Crises and emergencies lead to heightened risks. It is all a matter of degree. There is no standard definition of an emergency or crisis.

- **Is Any Help Better Than No Help?**—Situations in which any help is better than none raise interesting issues. If one can do more good than harm, then do it (Richards, 2002).

- **Suicide Risk**—The traditional scientific literature (Schneidman, Farberow, & Litman, 1983) thoroughly covers protocols for dealing with suicide risk and suicide assessment by telephone. Mental health professionals are expected to do what they can to prevent suicide.

- **Terminating the Professional Relationship**—Such an event should be discussed with the client prior to implementation. A policy or protocol should be available and spelled out for client at intake as part of the agreement client signs. Common sense should rule.

- **Standards of Care**—In the United States, state law defines the standards of care. Basically, a standard of care relates to the pattern of practice the profession generally accepts as reasonable under the relevant circumstances. Over time, some states have replaced local standards (local practices—e.g., early or current rural practices and approaches and beliefs) with state standards; other states have replaced their own state standards with national standards. Used in a legal context, a standard of care is typically established for a specific situation at a specific time. Online mental health practice has the potential to change how standards of care are viewed and to accelerate the evolution of national standards. The impact of the collaborative practice model using two-way video has been demonstrated to significantly change the diagnosis and treatment of problems and the management of clients/patients seeking care (Brown-Connolly, 2001b, p.37). The impact on outcomes and changes in standards of care will develop over time. The expectation is that collaborative practice would, in appropriate circumstances, tend to improve the quality of care.

- **Practice and Treatment Guidelines**—There is no doubt that various groups concerned with telehealth technologies will rapidly develop both practice and treatment guidelines including:
  o Administrative Standards
  o Technical Standards
  o Transmission
  o Security
  o Encryption
  o Passwords
  o Virtual Private Networks

As teleconferencing and teletherapy/counseling and other forms of tele-mental health and technologies become more common and cost-effective, it is critical that practitioners utilizing such methods thoroughly investigate the issues associated with establishing and implementing such approaches. In our rural areas of the west and other related geographical areas, they have great potential to provide mental/behavioral health and related services, including conferences, workshops, consultation and other approaches that remain difficult to access directly, especially in rural communities which may often be separated by long distances to include travel expenses, overnight stays, meals and other expenses. Providing such services in conjunction with community colleges and other resources can help our health care practitioners and others reach out to our rural areas. One example of where such an approach can be very helpful will be among our Veterans in rural communities. Well thought-out and planned approaches for implementation and maintenance and sustainability can make it work effectively.

Chapter 12: Traumatic Brain Injury: An Overview

Traumatic Brain Injury (TBI)

One area that we have not presented information about thus far in this volume is Traumatic Brain Injury (TBI). TBI is defined as damage to the brain resulting from external mechanical force, such as rapid acceleration or deceleration, blast waves, or penetration by a projectile. Brain function is temporarily or permanently impaired and structural damage may or may not be detectable with current technology. TBI can be caused by falls, vehicle accidents, and violence. It can cause a host of physical, cognitive, social, emotional and behavioral effects and outcomes can range from complete recovery to permanent disability or death. This particular chapter is not intended to present an in-depth explanation or review of research on TBI. However, a broad background and information about TBI is the goal of this chapter. There have been a large number of studies and reports that deal with various forms of TBI. In order to effectively review and present the information included in these will entail an additional volume that is planned for the near future. A brief reference list of a number of these relevant articles, research studies and reports is included at the end of this chapter. It is intended as a summary, not an extensive review. The references are provided for those who have an interest in looking deeper into the subject. A future volume is planned that is expected to address this topic in further detail. At least one presentation is also expected to be included in the next Conference of the Rocky Mountain Region Disaster Mental Health Institute in 2013.

Demystifying TBI: History and Background

In the Edwin Smith Papyrus (1650-1550 BC), head injury is referred to in ancient myths dating back before recorded history. Skulls that have been found in battleground graves with holes drilled over fracture lines suggest that trepanation may have been used to treat TBI in ancient times.

Ancient Mesopotamians and other cultures (e.g., ancient Inca burial sites) knew of head injury and some of its effects, including seizures, paralysis, and loss of sight, hearing or speech. The Edwin Smith Papyrus describes various head injuries and symptoms and classifies them based on their presentation and tractability. Hippocrates and other ancient Greek physicians understood the brain to be the center of thought, most likely due to their experience with head trauma. Surgeons during the Middle Ages and Renaissance continued to use the practice of trepanation for head injury. In the Middle Ages, physicians further described head injury symptoms and the term concussion became more widespread. Concussion symptoms were first described systematically in the 16th century by Berengario da Carpi.

In the 18th century intracranial pressure was suggested as the cause of pathology rather than skull damage following TBI. This was confirmed toward the end of the 19th century when it was proposed that opening the skull to relieve pressure be the treatment.

During the 19th century it was noted that TBI was related to the development of psychosis and a debate began about whether post-concussion syndrome was due to a disturbance of the brain tissue or psychological factors, a debate that continues today.

Personality Change: The Case of Phineas Gage

Phineas Gage is a name that one encounters in the literature often. He is perhaps the first reported case of personality change following brain injury. He survived an accident in which a large iron rod was driven through his head, destroying one or both of his frontal lobes. Since his case was first reported, numerous cases of personality change following brain injury have been reported.

During the 20th century the advancement of technologies improved treatment and diagnosis (e.g., the development of imaging tools including CT and MRI, and, in the 21st century, diffusion tensor imaging (DTI). The introduction of intracranial pressure monitoring in the 1950s has been credited with beginning the "modern era" of head injury. Prior to the 20th century, mortality rates of TBI were high. Rehabilitation was uncommon. Improvements in care developed during World War I reduced the death rate and made rehabilitation possible. Facilities dedicated to TBI rehabilitation were initially established during that period. Explosives used in World War I caused many blast injuries. The large number of TBIs resulting permitted researchers to learn about localization of brain functions. Blast-related injuries today are common problems in returning veterans from Iraq & Afghanistan. Research suggests that the symptoms of such TBIs are generally the same as those of TBIs involving a physical blow to the head.

During the 1970s, awareness of TBI as a public health problem developed. A great deal of progress has been made in brain trauma research since the 1970s. Examples include the discovery of primary and secondary brain injury. During the 1990s, the development and dissemination of standardized guidelines for treatment of TBI, with protocols for a range of issues, including drugs and management of intracranial pressure, has continued to develop. Research since the early 1990s has improved TBI survival. That decade has become known as the "Decade of the Brain" due to the advances made in brain research.

Some of our deployed service members have sustained attacks from explosions or blasts by rocket-propelled grenades, improvised explosive devices (IEDs) and land mines. Since October 24, 2008, the Veterans Administration has increased the disability rating for TBI Veterans. Depending on the extent of the injury, Veterans are eligible for up to 100% disability rating. TBI is the signature injury of the wars in Iraq and Afghanistan. One promising technology for treatment is called activation database guided EEG biofeedback. It has been documented to help return a TBIs auditory memory ability to above that of the performance of control groups.

Brief Overview and Background

Brain injuries are usually classified into mild, moderate, and severe categories. The Glasgow Coma Scale (GCS), is the most commonly used system for classifying TBI severity. It is used to grade a person's level of consciousness on a scale of 3–15 based on verbal, motor, and eye-opening reactions to stimuli. Generally a TBI with a GCS of 13 or above is considered mild, 9–12 moderate, and 8 or below severe. However, the GCS grading system is limited in its ability to predict outcomes. Therefore, other classification systems are also used to help determine severity. A current model developed by the Department of Defense and Department of Veterans Affairs uses all three criteria of GCS following resuscitation, duration of post-traumatic amnesia (PTA), and loss of consciousness (LOC). It also has been proposed to use changes that are visible on neuroimaging, such as swelling, focal lesions, or diffuse injury as a method of classification. Grading scales also exist to classify the severity of mild TBI, commonly called concussion. These use duration of LOC, PTA, and other concussion symptoms.

There are also Systems that exist to classify TBI by its pathological features. Damage from TBI can be focal or diffuse. It can be confined to specific areas or distributed in a more general manner. However, it is common for both types of injury to exist in a given case.

Diffuse injury manifests with little apparent damage in neuroimaging studies. In the early 2000s, researchers discovered that diffusion tensor imaging (DTI), a way of processing MRI images, was an effective tool for displaying the extent of diffuse axonal injury. Types of injuries considered diffuse include edema (swelling) and diffuse axonal injury, which is widespread damage to axons including white matter tracts and projections to the cortex. Types of injuries considered diffuse include concussion and diffuse axonal injury, widespread damage to axons in areas including white matter and the cerebral hemispheres.

Focal injuries often produce symptoms related to the functions of the damaged area. Research shows that the most common areas that have focal lesions in non-penetrating traumatic brain injury are the orbitofrontal cortex (the lower surface of the frontal lobes) and the anterior temporal lobes. These are areas involved in social behavior, emotion regulation, olfaction, and decision-making. They are the common social/emotional and judgment deficits that follow moderate-severe TBI. Symptoms such as hemiparesis or aphasia may also occur when less commonly affected areas such as motor or language areas are damaged.

Cerebral laceration (a type of focal injury), occurs when the tissue is cut or torn. Tearing is common in the orbitofrontal cortex in

particular. This is because of bony protrusions on the interior skull ridge above the eyes. A similar injury, cerebral contusion (bruising of brain tissue), involves blood mixed among tissue. In contrast, intracranial hemorrhage involves bleeding that is not mixed with tissue.

Hematomas, (which are also focal lesions), are collections of blood in or around the brain that can result from hemorrhage. Intracerebral hemorrhage, with bleeding in the brain tissue itself, is an intra-axial lesion. Extra-axial lesions include epidural hematoma, subdural hematoma, subarachnoid hemorrhage, and intraventricular hemorrhage. Epidural hematoma involves bleeding into the area between the skull and the dura mater, the outermost of the three membranes surrounding the brain. In subdural hematoma, bleeding occurs between the dura and the arachnoid mater. Subarachnoid hemorrhage involves bleeding into the space between the arachnoid membrane and the pia mater. Intraventricular hemorrhage occurs when there is bleeding in the ventricles.

Signs and symptoms: What to look for?

There are some signs or indications that generally suggest TBI. If these are noticed, it is important that they be checked. Referral to a specialist in this area is suggested in order to evaluate the situation. The following are some signs to be aware of and look for if exposed to a possible TBI event.

Unequal pupil size is a potential sign of a serious brain injury. Symptoms are dependent on the type of TBI (diffuse or focal) and the part of the brain that is affected. Unconsciousness tends to last longer for people with injuries on the left side of the brain than for those with injuries on the right. Symptoms are also dependent on the severity of the injury. With mild TBI, the person may remain conscious or may lose consciousness for a few seconds or minutes. Other symptoms of mild TBI include headache, vomiting, nausea, lack of motor coordination, dizziness, difficulty balancing, lightheadedness, blurred vision or tired eyes, ringing in the ears, bad taste in the mouth, fatigue or lethargy, and changes in sleep patterns. Cognitive and emotional symptoms include behavioral or mood changes, confusion, and trouble with memory, concentration, attention, or thinking. Mild TBI symptoms may also be present in moderate and severe injuries.

A person with moderate or severe TBI may have a headache that does not go away, repeated vomiting or nausea, convulsions, an inability to awaken, dilation of one or both pupils, slurred speech, aphasia (word-finding difficulties), dysarthria (muscle weakness that causes disordered speech), weakness or numbness in the limbs, loss of coordination, confusion, restlessness, or agitation. Common long-term symptoms of moderate to severe TBI include changes in appropriate social behavior, deficits in social judgment, and cognitive changes, especially problems with sustained attention, processing speed, and executive functioning. Alexithymia, a deficiency in identifying, understanding, processing, and describing emotions occurs in about two-thirds of individuals with TBI. Cognitive and social deficits have long-term consequences for the daily lives of people with moderate to severe TBI, but can be improved with appropriate rehabilitation.

When pressure within the skull (intracranial pressure—ICP) rises too high, it can be deadly. Signs of increased ICP include decreasing level of consciousness, paralysis or weakness on one side of the body, and a blown pupil (one that fails to constrict in response to light or is slow to do so). Cushing's triad, a slow heart rate with high blood pressure and respiratory depression is a classic manifestation of significantly raised ICP. Anisocoria (unequal pupil size) is another sign of serious TBI. Abnormal posturing, a characteristic positioning of the limbs caused by severe diffuse injury or high ICP, is an ominous sign.

What are Some Causes of TBI?

The most common causes of TBI in the U.S. include violence, transportation accidents, construction, and sports. Motor bikes are major causes, increasing in significance in developing countries as other causes reduce. A large number of traumatic brain injuries annually result from sports and recreation activities in the US. TBI is the third most common injury to result from child abuse. Abuse causes 19% of cases of pediatric brain trauma, and the death rate is higher among these cases. Domestic violence is another cause of TBI, as are work-related and industrial accidents. Firearms and blast injuries from explosions are other causes of TBI, which is the leading cause of death and disability in war zones. According to Representative Bill Pascrell (Democrat, NJ), TBI is "the signature injury of the wars in Iraq and Afghanistan." There is a promising technology called activation database guided EEG biofeedback which has been documented to return a TBI's auditory memory ability to above the control group's performance

Physical Forces

Ricochet of the brain within the skull may account for the coup-contrecoup phenomenon. The type, direction, intensity, and duration of forces all contribute to the characteristics and severity of TBI. Forces that may contribute to TBI include angular, rotational, shear, and translational forces. Even in the absence of an impact, significant acceleration or deceleration of the head can cause TBI. In most cases a combination of impact and acceleration is probably to blame. Forces involving the head striking or being struck by something, (termed contact or impact loading), are the cause of most focal injuries, and movement of the brain within the skull, (termed noncontact or inertial loading), usually causes diffuse injuries. The violent shaking of an infant that causes shaken baby syndrome commonly manifests as diffuse injury. In impact loading, the force sends shock waves through the skull and brain, resulting in tissue damage. Shock waves caused by penetrating injuries can also destroy tissue along the path of a projectile, compounding damage caused by the missile itself.

Damage may occur directly under the site of impact, or it may occur on the side opposite the impact (coup and contrecoup injury, respectively). When a moving object impacts the stationary head, coup injuries are typical, while contrecoup injuries are usually produced when the moving head strikes a stationary object.

Primary and secondary injuries

A large percentage of people killed by brain trauma do not die right away but rather days to weeks following the event. Rather than improving after being hospitalized, some TBI patients deteriorate. Primary brain injury (the damage that occurs at the moment of trauma when tissues and blood vessels are stretched, compressed, and torn) is not adequate to explain this deterioration. It is caused by secondary injury, (a complex set of cellular processes and biochemical cascades that occur in the minutes to days following the trauma). These secondary processes can dramatically worsen damage caused by primary injury and account for the greatest number of TBI deaths occurring in hospitals.

Secondary injury events include damage to the blood–brain barrier, release of factors that cause inflammation, free radical overload, excessive release of the neurotransmitter glutamate (excitotoxicity), influx of calcium and sodium ions into neurons, and dysfunction of mitochondria. Injured axons in the brain's white matter may separate from their cell bodies as a result of secondary injury, potentially killing those neurons. Other factors in secondary injury are changes in the blood flow to the brain; ischemia (insufficient blood flow); cerebral hypoxia (insufficient oxygen in the brain); cerebral edema (swelling of the brain); and raised intracranial pressure (the pressure within the skull). Intracranial pressure may rise due to swelling or a mass effect from a lesion, such as a hemorrhage. As a result, cerebral perfusion pressure (the pressure of blood flow in the brain) is reduced and ischemia results. When the pressure within the skull rises too high, it can cause brain death or herniation, in which parts of the brain are squeezed by structures in the skull. A particularly weak part of the skull that is vulnerable to damage causing extradural hematoma is the pterion, deep in which lies the middle meningeal artery which is easily damaged in fractures of the pterion. Since the pterion is so weak this type of injury can easily occur and can be secondary due to trauma to other parts of the skull where the impact forces spreads to the pterion.

Diagnosis

Neuroimaging helps in determining the diagnosis and prognosis and in deciding what treatments to give. The preferred radiologic test in emergency settings is computed tomography (CT). It is quick, accurate, and widely available. Follow-up CT scans may be performed later to determine whether the injury has progressed. Magnetic resonance imaging (MRI) can show more detail than CT, and adds information about expected outcome in the long term. It is more useful than CT for detecting injury characteristics such as diffuse axonal injury in the longer term. However, MRI is not used in the emergency setting for reasons including its relative inefficacy in detecting bleeds and fractures, its lengthy acquisition of images, the inaccessibility of the patient in the machine, and its incompatibility with metal items used in emergency care.

Other techniques may be used to confirm a particular diagnosis. X-rays are still used for head trauma, but evidence suggests they are not useful. Head injuries are either so mild that they do not need imaging or severe enough to merit the more accurate CT. Angiography may be used to detect blood vessel pathology when risk factors such as penetrating head trauma are involved. Functional imaging can measure cerebral blood flow or metabolism, inferring neuronal activity in specific regions and potentially helping to predict outcome. Electroencephalography and transcranial doppler may also be used. The most sensitive physical measure to date is the quantitative EEG which has documented an 80% to 100% ability in discriminating between normal and traumatic brain injured subjects.

Neuropsychological assessment can be performed to evaluate the long-term cognitive sequelae and to aid in the planning of rehabilitation. Instruments range from short measures of general mental functioning to complete batteries formed of different domain-specific tests.

Treatment

It is important to begin emergency treatment within the so-called "golden hour" following the injury. People with moderate to severe injuries are likely to receive treatment in an intensive care unit followed by a neurosurgical ward. Treatment depends on the recovery stage of the patient. In the acute stage the primary goal of medical personnel is to stabilize the patient and focus on preventing further injury because little can be done to reverse the initial damage caused by trauma. Rehabilitation is the main treatment for the subacute and chronic stages of recovery. International clinical guidelines have been proposed with the goal of guiding decisions in TBI treatment, as defined by an authoritative examination of current evidence.

Recent clinical and laboratory research by neurosurgeon Julian Bailes, M.D., and his colleagues from West Virginia University, has resulted in papers showing that dietary supplementation with omega-3 DHA (Docosahexaenoic Acid) offers protection against the biochemical brain damage that occurs after a traumatic injury. Rats given DHA prior to induced brain injuries suffered smaller increases in two key markers for brain damage (APP and caspase-3), as compared with rats given no DHA. "The potential for DHA to provide prophylactic benefit to the brain against traumatic injury appears promising and requires further investigation. The essential concept of daily dietary supplementation with DHA, so that those at significant risk may be preloaded to provide protection against the acute effects of TBI, has tremendous public health implications."

Certain facilities are equipped to handle TBI better than others. Initial measures include transporting patients to an appropriate treatment center. Both during transport and in hospital primary concerns include ensuring proper oxygen supply, maintaining adequate cerebral blood flow, and controlling raised intracranial pressure (ICP), since high ICP deprives the brain of badly needed blood flow and can cause deadly brain herniation. Other methods to prevent damage include management of other injuries and prevention of seizures. Neuroimaging is helpful but not flawless in detecting raised ICP. A more accurate way suggested to measure ICP is to place a catheter into a ventricle of the brain, which has the added benefit of allowing cerebrospinal fluid to drain, releasing pressure in the skull. Treatment of raised ICP may be as simple as tilting the patient's bed and straightening the head to promote blood flow through the veins of the neck. Sedatives, analgesics and paralytic agents are often used. Hypertonic saline can improve ICP by reducing the amount of cerebral water (swelling). However, it is used with caution to avoid electrolyte imbalances or heart failure. Mannitol, an osmotic diuretic, was also studied for this purpose. However, such studies have been inconclusive. Diuretics, drugs that increase urine output to reduce excessive fluid in the system, may be used to treat high intracranial pressures, but may cause hypovolemia (insufficient blood volume). Hyperventilation (larger and/or faster breaths) reduces carbon dioxide levels and causes blood vessels to constrict. This decreases blood flow to the brain and reduces ICP, but it potentially causes ischemia, and is therefore used only in the short term.

Endotracheal intubation and mechanical ventilation may be used to ensure proper oxygen supply and provide a secure airway.

Hypotension (low blood pressure), which has a devastating outcome in TBI, can be prevented by giving intravenous fluids to maintain a normal blood pressure. Failing to maintain blood pressure can result in inadequate blood flow to the brain. Blood pressure may be kept at an artificially high level under controlled conditions by infusion of norepinephrine or similar drugs. This helps maintain cerebral perfusion. Body temperature is carefully regulated because increased temperature raises the brain's metabolic needs. It potentially deprives it of nutrients. Seizures are common. While they can be treated with benzodiazepines, these drugs are used carefully because they can depress breathing and lower blood pressure. TBI patients are more susceptible to side effects and may react adversely or be inordinately sensitive to some pharmacological agents. During treatment monitoring continues for signs of deterioration such as a decreasing level of consciousness. Traumatic brain injury may cause a range of serious coincidental complications which include cardiac arrhythmias and neurogenic pulmonary edema. These conditions must be adequately treated and stabilized as part of the core care for these patients.

Surgery can be performed on mass lesions or to eliminate objects that have penetrated the brain. Mass lesions such as contusions or hematomas causing a significant mass effect (shift of intracranial structures) are considered emergencies and are removed surgically. For intracranial hematomas, the collected blood may be removed using suction or forceps or it may be floated off with water. Surgeons look for hemorrhaging blood vessels and seek to control bleeding. In penetrating brain injury, damaged tissue is surgically debrided, and craniotomy may be needed. Craniotomy, in which part of the skull is removed, may be needed to remove pieces of fractured skull or objects embedded in the brain. Decompressive craniectomy (DC) is performed routinely in the very short period following TBI during operations to treat hematomas; part of the skull is removed temporarily (primary DC). DC performed hours or days after TBI in order to control high intracranial pressures (secondary DC) has not been shown to improve outcome in some trials and may be associated with severe side effects. It is important that family and individuals consult appropriate professionals concerning the above information to help understand treatments.

Research

No medication to halt the progression of secondary injury exists. However, the variety of pathological events presents opportunities to find treatments that interfere with the damage processes. Neuroprotection, methods to halt or mitigate secondary injury, have been the subject of great interest for their ability to limit the damage that follows TBI. However, clinical trials to test agents that could halt these cellular mechanisms have met largely with failure. For example, interest existed in hypothermia, cooling the injured brain to limit TBI damage. Clinical trials, however, have shown that it is not useful in the treatment of TBI. In addition, drugs such as NMDA receptor antagonists to halt neurochemical cascades such as excitotoxicity have shown promise in animal trials but failed in clinical trials. These failures could be due to factors including faults in the trials' design or in the insufficiency of a single agent to prevent the array of injury processes involved in secondary injury. Recent research has looked into monitoring brain metabolism for ischaemia, in particular the parameters of glucose, glycerol, and glutamate through microdialysis.

Developments in technologies may provide doctors with valuable medical information. For example, work has been done to design a device to monitor oxygenation that could be attached to a probe placed into the brain—such probes are currently used to monitor ICP. Research is also planned to clarify factors correlated to outcome in TBI and to determine in which cases it is best to perform CT scans and surgical procedures.

Hyperbaric oxygen therapy (HBO) has been evaluated as an adjunctive treatment following TBI, concluding a Cochrane review stating that its use could not be justified. HBO for TBI has remained controversial as studies have looked for improvement mechanisms, and further evidence shows that it may have potential as a treatment

Rehabilitation

Physical therapy will commonly include muscle strength exercise. Once medically stable, patients may be transferred to a subacute rehabilitation unit of the medical center or to an independent rehabilitation hospital. Rehabilitation works to improve independent function at home and in society and to help adapt to disabilities. It has demonstrated its general effectiveness when conducted by a team of health professionals who specialize in head trauma. As true for any patient with neurologic deficits, a multidisciplinary approach is key to optimizing outcome. Physiatrists or neurologists are likely to be the key medical staff involved. Depending on the patient, doctors of other medical specialties may also be helpful. Allied health professions such as physiotherapy, speech and language therapy, cognitive rehabilitation therapy, and occupational therapy will be essential to assess function and design the rehabilitation activities for each patient. Treatment of neuropsychiatric symptoms such as emotional distress and clinical depression may involve mental health professionals such as therapists, psychologists, and psychiatrists, while neuropsychologists can help to evaluate and manage cognitive deficits. Following discharge from the inpatient rehabilitation treatment unit, care may be given on an outpatient basis. Community-based rehabilitation will be required for a high proportion of patients, including vocational rehabilitation. This supportive employment matches job demands to the worker's abilities. People with TBI who cannot live independently or with family may require care in supported living facilities such as group homes. Respite care, including day centers and leisure facilities for the disabled, offers time off for caregivers, and activities for people with TBI.

Pharmacological treatment can help to manage psychiatric or behavioral problems. Medication is also used to control post-traumatic epilepsy. However, preventive use of anti-epileptics is not recommended. In those cases where the person is bedridden due to a reduction of consciousness, has to remain in a wheelchair because of mobility problems, or has any other problem heavily impacting self-caring capacities, caregiving and nursing are critical. The most effective research documented intervention approach is the activation database guided EEG biofeedback approach which has shown significant improvements in memory abilities of the TBI individual which are far superior to traditional approaches (strategies, computers, medication intervention). Gains of 2.61 standard deviations have been documented. The TBI's auditory memory ability was superior to the control group following the treatment.

What Is the Prognosis for TBI Individuals?

Prognosis worsens with the severity of injury. Most TBIs are mild and do not cause permanent or long-term disability. However, all severity levels of TBI have potential to cause significant, long-lasting disability. Permanent disability is thought to occur in 10% of mild injuries, 66% of moderate injuries, and 100% of severe injuries. Most mild TBI is completely resolved within three weeks. Almost all people with mild TBI are able to live independently and return to the jobs they had before the injury. A portion have mild cognitive and social impairments. Over 90% of people with moderate TBI are able to live independently. A portion require assistance in areas such as physical abilities, employment, and financial managing. Most people with severe closed head injury either die or recover enough to live independently. Middle ground is less common. Coma, as it is closely related to severity, is a strong predictor of poor outcome.

Prognosis differs depending on the severity and location of the lesion, and access to immediate, specialized acute management.

Subarachnoid hemorrhage approximately doubles mortality. Subdural hematoma is associated with worse outcome and increased mortality, while people with epidural hematoma are expected to have a good outcome if they receive surgery quickly. Diffuse axonal injury may be associated with coma when severe, and poor outcome. Following the acute stage, prognosis is strongly influenced by the patient's involvement in activity that promotes recovery, which for most patients requires access to a specialized, intensive rehabilitation service.

Medical complications are associated with a bad prognosis. Examples include hypotension (low blood pressure), hypoxia (low blood oxygen saturation), lower cerebral perfusion pressures and longer times spent with high intracranial pressures. Patient characteristics also influence prognosis. Factors thought to worsen it include abuse of substances such as illicit drugs and alcohol and age over sixty or under two years (in children, younger age at time of injury may be associated with a slower recovery of some abilities).

Complications of Traumatic Brain Injury

The relative risk of post-traumatic seizures increases with the severity of traumatic brain injury. Improvement of neurological function usually occurs for two or more years after the trauma. For many years it was believed that recovery was fastest during the first six months. However, there is no evidence to support this. It may be related to services commonly being withdrawn after this period rather than any physiological limitation to further progress. Children recover better in the immediate time frame and improve for longer periods.

Complications are distinct medical problems that may arise as a result of the TBI. Results of traumatic brain injury vary widely in type and duration. They include physical, cognitive, emotional, and behavioral complications. TBI can cause prolonged or permanent effects on consciousness, such as coma, brain death, persistent vegetative state (in which patients are unable to achieve a state of alertness to interact with their surroundings), and minimally conscious state (in which patients show minimal signs of being aware of self or environment). Lying still for long periods can cause complications including pressure sores, pneumonia or other infections, progressive multiple organ failure, and deep venous thrombosis, which can cause pulmonary embolism. Infections that can follow skull fractures and penetrating injuries include meningitis and abscesses. Complications involving the blood vessels include vasospasm, in which vessels constrict and restrict blood flow, the formation of aneurysms, in which the side of a vessel weakens and balloons out, and stroke.

Movement disorders that may develop after TBI include tremor, ataxia (uncoordinated muscle movements), myoclonus (shock-like contractions of muscles), and loss of movement range and control (in particular with a loss of movement repertoire). The risk of post-traumatic seizures increases with severity of trauma and is particularly elevated with certain types of brain trauma such as cerebral contusions or hematomas. People with early seizures, those occurring within a week of injury, have an increased risk of post-traumatic epilepsy (recurrent seizures occurring more than a week after the initial trauma). People may lose or experience altered vision, hearing, or smell.

Hormonal disturbances may occur secondary to hypopituitarism, occurring immediately or years after injury. Development of diabetes insipidus or an electrolyte abnormality acutely after injury indicate need for endocrinological workup. Signs and symptoms of hypopituitarism may develop and be screened for in adults with moderate TBI and in mild TBI with imaging abnormalities. Children with moderate to severe head injury may also develop hypopituitarism. Screening should take place 3 to 6 months, and 12 months after injury, but problems may occur more remotely.

Cognitive deficits that can follow TBI include impaired attention; disrupted insight, judgment, and thought; reduced processing speed; distractibility; and deficits in executive functions such as abstract reasoning, planning, problem-solving, and multitasking. Memory loss, the most common cognitive impairment among head-injured people, occurs in 20–79% of people with closed head trauma, depending on severity. People who have suffered TBI may also have difficulty with understanding or producing spoken or written language, or with more subtle aspects of communication such as body language. Post-concussion syndrome, a set of lasting symptoms experienced after mild TBI, can include physical, cognitive, emotional and behavioral problems such as headaches, dizziness, difficulty concentrating, and depression. Multiple TBIs may have a cumulative effect. A young person who receives a second concussion before symptoms from another one have healed may be at risk for developing a very rare but deadly condition called *second-impact syndrome*, in which the brain swells catastrophically after even a mild blow, with debilitating or deadly results. About one in five career boxers is affected by chronic traumatic brain injury (CTBI), which causes cognitive, behavioral, and physical impairments. *Dementia pugilistica*, the severe form of CTBI, affects primarily career boxers years after a boxing career. It commonly manifests as dementia, memory problems, and parkinsonism (tremors and lack of coordination).

TBI may cause emotional, social, or behavioral problems and changes in personality. These may include emotional instability, depression, anxiety, hypomania, mania, apathy, irritability, problems with social judgment, and impaired conversational skills. TBI appears to predispose survivors to psychiatric disorders including obsessive compulsive disorder, substance abuse, dysthymia, clinical depression, bipolar disorder, and anxiety disorders. In patients who have depression after TBI, suicidal ideation is not uncommon. The suicide rate among these persons is increased. Social and behavioral symptoms that can follow TBI include disinhibition, inability to control anger, impulsiveness, lack of initiative, inappropriate sexual activity, poor social judgment, and changes in personality.

TBI also has a substantial impact on the functioning of family systems. Caregiving family members and TBI survivors often significantly alter their familial roles and responsibilities following injury, creating significant change and strain on a family system. Typical challenges identified by families recovering from TBI include: frustration and impatience with one another, loss of former lives and relationships, difficulty setting reasonable goals, inability to effectively solve problems as a family, increased level of stress and household tension, changes in emotional dynamics, and overwhelming desire to return to pre-injury status. In addition, families may exhibit less effective functioning in areas including coping, problem solving and communication. Psychoeducation and counseling models have been demonstrated to be effective in minimizing family disruption.

Demographics

TBI is present in 85% of traumatically injured children, either alone or with other injuries. The greatest number of TBIs occur in people aged 15–24. Because TBI is more common in young people, its costs to society are high due to the loss of productive years to death and disability. The age groups most at risk for TBI are children ages five to nine and adults over age 80, and the highest rates of death and hospitalization due to TBI are in people over age 65. The incidence of fall-related TBI in First World countries is increasing as the population ages; thus the median age of people with head injuries has increased. Regardless of age, TBI rates are higher in males. Men suffer twice as many TBIs as women do and have a fourfold risk of fatal head injury, and males account for two thirds of childhood and adolescent head trauma. However, when matched for severity of injury, women appear to fare more poorly than men. Socioeconomic status also appears to affect TBI rates; people with lower levels of education and employment and lower socioeconomic status are at greater risk.

Causes of TBI fatalities in the US and Other Countries

TBI is a leading cause of death and disability around the globe and presents a major worldwide social, economic, and health problem. It is the number one cause of coma. It plays the leading role in disability due to trauma and is the leading cause of brain damage in children and young adults. In Europe it is responsible for more years of disability than any other cause. It also plays a significant role in half of trauma deaths. Findings on the frequency of each level of severity vary based on the definitions and methods used in studies. A World Health Organization study estimated that between 70 and 90% of head injuries that receive treatment are mild, and a US study found that moderate and severe injuries each account for 10% of TBIs, with the rest mild.

The incidence of TBI varies by age, gender, region and other factors. Findings of incidence and prevalence in epidemiological studies vary based on such factors as which grades of severity are included, whether deaths are included, whether the study is restricted to hospitalized people, and the study's location. The annual incidence of mild TBI is difficult to determine but may be 100–600 people per 100,000. In the US, the mortality (death rate) rate is estimated to be 21% by 30 days after TBI. A study on Iraq War veterans found that severe TBI carries a mortality of 30–50%. Deaths have declined due to improved treatments and systems for managing trauma in societies wealthy enough to provide modern emergency and neurosurgical services. The fraction of those who die after being hospitalized with TBI fell from almost half in the 1970s to about a quarter at the beginning of the 21st century. This decline in mortality has led to a concomitant increase in the number of people living with disabilities that result from TBI. Biological, clinical, and demographic factors contribute to the likelihood that an injury will be fatal. In addition, outcome depends heavily on the cause of head injury. In the US, patients with fall-related TBIs have an 89% survival rate, while only 9% of patients with firearm-related TBIs survive.[In the US, firearms are the most common cause of fatal TBI, followed by vehicle accidents and then falls. Of deaths from firearms, 75% are considered to be suicides.

The incidence of TBI is increasing globally, due largely to an increase in motor vehicle use in low- and middle-income countries. In developing countries, automobile use has increased faster than safety infrastructure could be introduced. In contrast, vehicle safety laws have decreased rates of TBI in high-income countries, which have seen decreases in traffic-related TBI since the 1970s. Each year in the United States about two million people suffer a TBI and about 500,000 are hospitalized. The yearly incidence of TBI is estimated at 180–250 per 100,000 people in the US,

281 per 100,000 in France, 361 per 100,000 in South Africa, 322 per 100,000 in Australia, and 430 per 100,000 in England. In the European Union the yearly aggregate incidence of TBI hospitalizations and fatalities is estimated at 235 per 100,000.

References

The following references were drawn on for the chapter above. Rather than following the traditional method of the names and dates within the text, they are listed below. This was done for this chapter only because there is an extremely large number of relevant references pertaining to this chapter. It is also done in order to make them more available to those seeking further information. As it is planned to have a more detailed volume that will include more specific information on the topic, this chapter is meant only as an overview and general introduction to the topic of Traumatic Brain Injury.

Rehman T, Ali R, Tawil I, Yonas H (2008). "Rapid progression of traumatic bifrontal contusions to transtentorial herniation: A case report". *Cases Journal* 1 (1): 203. http://www.casesjournal.com/content/1/1/203.

Maas AI, Stocchetti N, Bullock R (August 2008). "Moderate and severe traumatic brain injury in adults". *Lancet Neurology* 7 (8): 728–41.

Parikh S, Koch M, Narayan RK (2007). "Traumatic brain injury". International Anesthesiology Clinics 45 (3): 119–35.

Chapman SB, Levin HS, Lawyer SL (1999). "Communication problems resulting from brain injury in children: Special issues of assessment and management". In McDonald S, Togher L, Code C. Communication Disorders Following Traumatic Brain Injury. East Sussex: Psychology Press. pp. 235–36. ISBN 0-86377-724-4. http://books.google.com/?id=klwVAAAAIAAJ&pg=PA236&dq=non-traumatic+%22Acquired+brain+injury. Retrieved 2008-11-13.

Collins C, Dean J (2002). "Acquired brain injury". In Turner A, Foster M, Johnson SE. Occupational Therapy and Physical Dysfunction: Principles, Skills and Practice. Edinburgh: Churchill Livingstone. pp. 395–96. ISBN 0-443-06224-2. http://books.google.com/?id=z2sA3mnG_zUC&pg=PA395&dq=non-traumatic+%22Acquired+brain+injury. Retrieved 2008-11-13.

Blissitt PA (2006). "Care of the critically ill patient with penetrating head injury". *Critical Care Nursing Clinics of North America* 18 (3): 321–32.

Hannay HJ, Howieson DB, Loring DW, Fischer JS, Lezak MD (2004). "Neuropathology for neuropsychologists". In Lezak MD, Howieson DB, Loring DW. Neuropsychological Assessment. Oxford [Oxfordshire]: Oxford University Press. pp. 158–62. ISBN 0-19-511121-4.

Jennett B (May 1998). "Epidemiology of head injury". Archives of Disease in Childhood 78 (5): 403–06. doi:10.1136/adc.78.5.403. PMC 1717568. PMID 9659083. http://adc.bmj.com/cgi/content/full/78/5/403.

McCaffrey RJ (1997). "Special issues in the evaluation of mild traumatic brain injury". The Practice of Forensic Neuropsychology: Meeting Challenges in the Courtroom. New York: Plenum Press. pp. 71–75. ISBN 0-306-45256-1.

LaPlaca et al. (2007). p.16

Weber JT, Maas AIR (2007). Weber JT. ed. Neurotrauma: New Insights Into Pathology and Treatment. Amsterdam: Academic Press. p. xi. ISBN 0-444-53017-7. http://books.google.com/?id=FyzEQPKUuPcC&pg=PP1&dq=neuro trauma. Retrieved 2008-11-12.

Saatman KE, Duhaime AC Workshop Scientific Team Advisory Panel Members et al. (2008). "Classification of traumatic brain injury for targeted therapies". *Journal of Neurotrauma* 25 (7): 719–38. //www.ncbi.nlm.nih.gov/pmc/articles/PMC2721779/.

Department of Defense and Department of Veterans Affairs (2008). "Traumatic Brain Injury Task Force". http://www.cdc.gov/nchs/data/icd9/Sep08TBI.pdf.

Marion (1999). p.4.

Valadka AB (2004). "Injury to the cranium". In Moore EJ, Feliciano DV, Mattox KL. Trauma. New York: McGraw-Hill, Medical Pub. Division. pp. 385–406. ISBN 0-07-137069-2. http://books.google.com/?id=VgizxQg-8QQC&pg=PA385. Retrieved 2008–08–15.

Hayden MG, Jandial R, Duenas HA, Mahajan R, Levy M (2007).

"Pediatric concussions in sports: A simple and rapid assessment tool for concussive injury in children and adults". *Child's Nervous System* 23 (4): 431–435.

Seidenwurm DI (2007). "Introduction to brain imaging". In Brant WE, Helms CA. Fundamentals of Diagnostic Radiology. Philadelphia: Lippincott, Williams & Wilkins. pp. 53–55. ISBN 0-7817-6135-2. http://books.google.com/?id=Sossht2t5XwC&pg=PA53&lpg=PA53 &dq=extra-axial+intra-axial. Retrieved 2008-11-17.

Smith DH, Meaney DF, Shull WH (2003). "Diffuse axonal injury in head trauma". *Journal of Head Trauma Rehabilitation* 18 (4): 307–16.

Granacher (2007). p.32.

Kraus, M.F.; Susmaras,T., Caughlin, B.P., Walker, C.J., Sweeney, J.A., & Little, D.M. (2007). "White matter integrity and cognition in chronic traumatic brain injury: A diffusion tensor imaging study". *Brain* 130 (10): 2508–2519.

Kumar, R.; Husain M, Gupta RK, Hasan KM, Haris M, Agarwal AK, Pandey CM, Narayana PA (Feb 2009). "Serial changes in the white matter diffusion tensor imaging metrics in moderate traumatic brain injury and correlation with neuro-cognitive function". *Journal of Neurotrauma* 26 (4): 481–495.

Melvin JW, Lighthall JW (2002). Nahum AM, Melvin JW. ed. Accidental Injury: Biomechanics and Prevention. Berlin: Springer. pp. 280–81. ISBN   0-387-98820-3. http://books.google.com/?id=Y4l5fopEI0EC&pg=PA280&dq=focal +diffuse+brain+injury. Retrieved 2008-11-15.

McCrea, M. (2007). Mild Traumatic Brain Injury and Postconcussion Syndrome: The New Evidence Base for Diagnosis and Treatment (American Academy of Clinical Neuropsychology Workshop Series). New York: Oxford University Press. ISBN 978-0-19-532829-5 ISBN 0195328299.

Mattson, A.J.; Levin, H.S. (1990). Frontal lobe dysfunction following closed head injury. A review of the literature. *Journal of Nervous & Mental Disorders* 178 (5): 282–291.

Bayly, P.V.; Cohen, T.S., Leister, E.P., Ajo, D., Leuthardt, E.C., & Genin, G.M. (2005). "Deformation of the human brain induced by mild acceleration". *Journal of Neurotrauma* 22 (8): 845–856.

www.ncbi.nlm.nih.gov/pmc/articles/PMC2377024/

Cummings, J.L. (1993). Frontal-subcortical circuits and human behavior. *Archives of Neurology* 50 (8): 873–880.

McDonald, S.; Flanagan, S., Rollins, J., & Kinch, J (2003). TASIT: A new clinical tool for assessing social perception after traumatic brain injury. *Journal of Head Trauma Rehabilitation* 18 (3): 219–238.

Basso A, Scarpa MT (December 1990). Traumatic aphasia in children and adults: a comparison of clinical features and evolution. *Cortex* 26 (4): 501–14.

Mohr JP, Weiss GH, Caveness WF et al. (December 1980). Language and motor disorders after penetrating head injury in Viet Nam. *Neurology* 30 (12): 1273–9.

Hardman JM, Manoukian A (2002). "Pathology of head trauma". Neuroimaging Clinics of North America 12 (2): 175–87, vii.

Barkley JM, Morales D, Hayman LA, Diaz-Marchan PJ (2006). "Static neuroimaging in the evaluation of TBI". In Zasler ND, Katz DI, Zafonte RD. Brain Injury Medicine: Principles and Practice. Demos Medical Publishing. pp. 140–43. ISBN 1-888799-93-5.

Ghajar J (September 2000). Traumatic brain injury. *Lancet* 356 (9233): 923–29.

Arlinghaus KA, Shoaib AM, Price TRP (2005). "Neuropsychiatric assessment". In Silver JM, McAllister TW, Yudofsky SC. Textbook Of Traumatic Brain Injury. Washington, DC: American Psychiatric Association. pp. 63–65. ISBN 1-58562-105-6.

"NINDS Traumatic Brain Injury Information Page". National Institute of Neurological Disorders and Stroke. 2008-09-15. http://www.ninds.nih.gov/disorders/tbi/tbi.htm. Retrieved 2008-10-27.

Kushner D (1998). Mild traumatic brain injury: Toward understanding manifestations and treatment. *Archives of Internal Medicine* 158 (15): 1617–24. http://archinte.highwire.org/cgi/content/full/158/15/1617

Stone, V.E.; Baron-Cohen, S., & Knight, R.T. (1998). Frontal lobe contributions to theory of mind. *Journal of Cognitive Neuroscience* 10 (5): 640–656.

Kim, E. (2002). Agitation, aggression, and disinhibition syndromes after traumatic brain injury. *NeuroRehabilitation* 17 (4): 297–310.

Busch, R. M.; McBride, A., Curtiss, G., & Vanderploeg, R. D. (2005). The components of executive functioning in traumatic brain injury. *Journal of Clinical and Experimental Neuropsychology* 27 (8): 1022–1032.

Ponsford, J.; K. Draper, and M. Schonberger (2008). Functional outcome 10 years after traumatic brain injury: its relationship with demographic, injury severity, and cognitive and emotional status. *Journal of the International Neuropsychological Society* 14 (2): 233–242.

Williams C, Wood RL (March 2010). Alexithymia and emotional empathy following traumatic brain injury. *J Clin Exp Neuropsychol* 32 (3): 259–67.

Milders, M.; Fuchs, S., & Crawford, J.R. (2003). Neuropsychological impairments and changes in emotional and social behaviour following severe traumatic brain injury. *Journal of Clinical & Experimental Neuropsychology* 25 (2): 157–172.

Ownsworth, T.; Fleming, J. (2005). The relative importance of metacognitive skills, emotional status, and executive function in psychosocial adjustment following acquired brain injury. *Journal of Head Trauma Rehabilitation* 20 (4): 315–332.

Dahlberg, C.A.; Cusick CP, Hawley LA, Newman JK, Morey CE, Harrison-Felix CL, & Whiteneck GG. (2007). Treatment efficacy of social communication skills training after traumatic brain injury: A randomized treatment and deferred treatment controlled trial. *Archives of Physical Medicine & Rehabilitation* 88 (12): 1561–1573.

Salomone JP, Frame SB (2004). "Prehospital care". In Moore EJ, Feliciano DV, Mattox KL. Trauma. New York: McGraw-Hill, Medical Pub. Division. pp. 117–8. ISBN 0-07-137069-2. http://books.google.com/?id=VgizxQg-8QQC&pg=PA545&dq=tracheobronchial. Retrieved 2008-08-15.

"Signs and Symptoms". Centers for Disease Control and Prevention, National Center for Injury Prevention and Control. 2007-07-07. http://www.cdc.gov/ncipc/tbi/Signs_and_Symptoms.htm. Retrieved 2008-10-27.

Faul, M (2010). "Traumatic Brain Injury in the United States: Emergency Department Visits, Hospitalizations, and Deaths, 2002-2006". National Center for Injury Prevention and Control, Centers for Disease Control and Prevention. http://www.cdc.gov/traumaticbraininjury/tbi_ed.html#3. Retrieved August 9, 2011.

Reilly P (2007). "The impact of neurotrauma on society: An international perspective". In Weber JT. Neurotrauma: New Insights Into Pathology and Treatment. Amsterdam: Academic Press. pp. 5–7. ISBN 0-444-53017-7. http://books.google.com/?id=FyzEQPKUuPcC&pg=PP1&dq=neuro trauma. Retrieved 2008-11-10.

"Traumatic brain injury". Centers for Disease Control and Prevention, National Center for Injury Prevention and Control. 2007. http://www.cdc.gov/ncipc/factsheets/tbi.htm. Retrieved 2008-10-28.

Granacher (2007). p.16.

Hunt JP, Weintraub SL, Wang YZ, Buetcher KJ (2004). "Kinematics of trauma". In Moore EJ, Feliciano DV, Mattox KL. Trauma. New York: McGraw-Hill, Medical Pub. Division. p. 153. ISBN 0-07-137069-2. http://books.google.com/?id=VgizxQg-8QQC&pg=PA545&dq=tracheobronchial. Retrieved 2008-08-15.

Elovic E, Zafonte R (2005). "Prevention". In Silver JM, McAllister TW, Yudofsky SC. Textbook of Traumatic Brain Injury. Washington, DC: American Psychiatric Association. p. 740. ISBN 1-58562-105-6.

Bay E, McLean SA (2007). Mild traumatic brain injury: An update for advanced practice nurses. *Journal of Neuroscience* Nursing 39 (1): 43–51.

Comper P, Bisschop SM, Carnide N et al. (2005). "A systematic review of treatments for mild traumatic brain injury". *Brain Injury* 19 (11): 863–880.

Champion, HR; Holcomb JB, Young LA (2009). Injuries from explosions. *Journal of Trauma* 66 (5): 1468–1476.

Park E, Bell JD, Baker AJ (April 2008). Traumatic brain injury: Can the consequences be stopped? *Canadian Medical Association Journal*

178   (9):   1163–70.
    //www.ncbi.nlm.nih.gov/pmc/articles/PMC2292762/.

"Pentagon Told Congress It's Studying Brain-Damage Therapy".
    ProPublica.   http://www.propublica.org/article/pentagon-told-
    congress-its-studying-brain-damage-therapy. Retrieved 2011-01-23.

Thornton, K. & Carmody, D. Efficacy of traumatic brain injury
    rehabilitation: interventions of QEEG-guided biofeedback,
    computers, strategies, and medications, *Applied Psychophysiology
    and Biofeedback*, 2008, (33) 2, 101-124.

Thornton, K. & Carmody, D. Traumatic brain injury rehabilitation:
    QEEG biofeedback treatment protocols, *Applied Psychophysiology
    and Biofeedback*, 2009, (34) 1, 59-68.

Shaw NA (2002). The neurophysiology of concussion. *Progress in
    Neurobiology*   67   (4):   281–344.   doi:10.1016/S0301-
    0082(02)00018-7. PMID 12207973.

American Academy of Pediatrics: Committee on Child Abuse and Neglect
    (July 2001). Shaken baby syndrome: Rotational cranial injuries.
    Technical   report.   *Pediatrics*   108   (1):   206–10.
    doi:10.1542/peds.108.1.206.
      http://pediatrics.aappublications.org/cgi/content/full/108/1/206.

Morrison AL, King TM, Korell MA, Smialek JE, Troncoso JC (1998).
    Acceleration-deceleration injuries to the brain in blunt force
    trauma. *American Journal of Forensic Medical Pathology* 19 (2):
    109–112.   doi:10.1097/00000433-199806000-00002.   PMID
    9662103.   http://meta.wkhealth.com/pt/pt-core/template-
    journal/lwwgateway/media/landingpage.htm?issn=0195-
    7910&volume=19&issue=2&spage=109.

Poirier MP (2003). Concussions: Assessment, management, and
    recommendations for return to activity (abstract). *Clinical Pediatric
    Emergency Medicine* 4 (3): 179–185.
    doi:10.1016/S1522-8401(03)00061-2.
    http://www.sciencedirect.com/science?_ob=ArticleURL&_udi=B75
    BD-49H1C2F-7.

Sauaia A., Moore F.A., Moore E.E. et al. (1995). Epidemiology of trauma
    deaths: A reassessment. *The Journal of Trauma* 38 (2): 185–93.
    doi:10.1097/00005373-199502000-00006.

Narayan RK, Michel ME, Ansell B et al. (2002). Clinical trials in head injury. *Journal of Neurotrauma* 19 (5): 503–57. www.ncbi.nlm.nih.gov/pmc/articles/PMC1462953/.

Xiong Y, Lee CP, Peterson PL (2000). "Mitochondrial dysfunction following traumatic brain injury". In Miller LP and Hayes RL, eds. Co-edited by Newcomb JK. Head Trauma: Basic, Preclinical, and Clinical Directions. New York: John Wiley and Sons, Inc. pp. 257–80. ISBN 0-471-36015-5.

Scalea T.M. (2005). "Does it matter how head injured patients are resuscitated?". In Valadka AB, Andrews BT. Neurotrauma: Evidence-based Answers to Common Questions. Thieme. pp. 3–4. ISBN 3-13-130781-1.

Morley E.J., Zehtabchi S. (September 2008). Mannitol for traumatic brain injury: Searching for the evidence. *Annals of Emergency Medicine* 52 (3): 298–300. doi:10.1016/j.annemergmed.2007.10.013.

Zink B.J. (2001). Traumatic brain injury outcome: Concepts for emergency care. *Annals of Emergency Medicine* 37 (3): 318–32. doi:10.1067/mem.2001.113505.

Barr R.M., Gean A.D., Le T.H. (2007). "Craniofacial trauma". In Brant WE, Helms CA. Fundamentals of Diagnostic Radiology. Philadelphia: Lippincott, Williams & Wilkins. p. 55. ISBN 0-7817-6135-2. http://books.google.com/?id=Sossht2t5XwC&pg=PA53&lpg=PA53&dq=extra-axial+intra-axial. Retrieved 2008-11-17.

Coles JP (July 2007). Imaging after brain injury. *British Journal of Anaesthesia* 99 (1): 49–60.

Thornton, K. Exploratory Analysis: Mild Head Injury, Discriminant Analysis with High Frequency Bands (32-64 Hz) under Attentional Activation Conditions & Does Time Heal?. Journal of Neurotherapy, Feb., 2000, 3 (3/4) 1-10.

Thornton, K., Exploratory Investigation into Mild Brain Injury and Discriminant Analysis with High Frequency Bands (32-64 Hz), Brain Injury, August, 1999, 477-488.

Liu BC, Ivers R, Norton R, Boufous S, Blows S, Lo SK (2008). Liu, Bette C. ed. Helmets for preventing injury in motorcycle riders. *Cochrane Database Syst Rev* (3): CD004333.

McCrory PR (August 2003). "Brain injury and heading in soccer". BMJ
    327 (7411): 351–52. doi:10.1136/bmj.327.7411.351. PMC
    1126775. PMID 12919964.
    http://www.bmj.com/cgi/content/full/327/7411/351?etoc

McIntosh AS, McCrory P (June 2005). Preventing head and neck injury.
    British Journal of Sports Medicine (Free registration required) 39
    (6): 314–18. http://bjsm.bmj.com/cgi/content/full/39/6/314.

Crooks CY, Zumsteg JM, Bell KR (November 2007). "Traumatic brain
    injury: A review of practice management and recent advances".
    Physical Medicine and Rehabilitation Clinics of North America 18
    (4): 681–710. doi:10.1016/j.pmr.2007.06.005.

Hahn RA, Bilukha O, Crosby A et al. (February 2005). "Firearms laws
    and the reduction of violence: a systematic review". Am J Prev Med
    28 (2 Suppl 1): 40–71.

Mills JD, Bailes JE, Sedney CL, Hutchins H, Sears B. (2011). Omega-3
    fatty acid supplementation and reduction of traumatic axonal
    injury in a rodent head injury model. *J Neurosurg.* 114 (1): 77–84.

Bailes JE, Mills JD. (2010). "Docosahexaenoic acid reduces traumatic
    axonal injury in a rodent head injury model". *J Neurotrauma.* 27
    (9): 1617–24.

Bailes JE, Mills JD, Hadley K. (2011). Dietary Supplementation with the
    Omega-3 Fatty Acid Docosahexaenoic Acid in Traumatic Brain
    Injury?. *Neurosurgery,* Epub ahead of print.

Kluger, Jeffrey. Dealing with Brain Injuries. *Time Magazine,* April 6,
    2009, p. 57. Online:
    http://www.time.com/time/magazine/article/0,9171,1887856,00.ht
    ml. Accessed: May 1, 2009

Office of Communications and Public Liaison (February 2002).
    "Traumatic brain injury: Hope through research". NIH Publication
    No. 02-2478. National Institute of Neurological Disorders and
    Stroke, National Institutes of Health.
    http://www.ninds.nih.gov/disorders/tbi/detail_tbi.htm. Retrieved
    2008-08-17.

Greenwald B.D., Burnett D.M., Miller M.A. (2003). "Congenital and
    acquired brain injury. 1. Brain injury: epidemiology and
    pathophysiology". *Archives of Physical Medicine and*

Rehabilitation 84 (3 Suppl 1): S3–7.

Moppett IK (July 2007). "Traumatic brain injury: Assessment, resuscitation and early management". British Journal of Anaesthesiology 99 (1): 18–31. http://bja.oxfordjournals.org/cgi/content/full/99/1/18. Moppet07

Gruen P (2002). Surgical management of head trauma. *Neuroimaging Clinics of North America* 12 (2): 339–43.

Cruz J, Minoja G, Okuchi K (2001). Improving clinical outcomes from acute subdural hematomas with the emergency preoperative administration of high doses of mannitol: A randomized trial. *Neurosurgery* 49 (4): 864–71.

Cruz J, Minoja G, Okuchi K (2002). Major clinical and physiological benefits of early high doses of mannitol for intraparenchymal temporal lobe hemorrhages with abnormal pupillary widening: A randomized trial. *Neurosurgery* 51 (3): 628–37; discussion 637–38.

Cruz J, Minoja G, Okuchi K, Facco E (March 2004). Successful use of the new high-dose mannitol treatment in patients with Glasgow Coma Scale scores of 3 and bilateral abnormal pupillary widening: A randomized trial. *Journal of Neurosurgery* 100 (3): 376–83.

Roberts I, Smith R, Evans S (February 2007). Doubts over head injury studies. *British Medical Journal* 334 (7590): 392–94. http://www.bmj.com/cgi/content/full/334/7590/392.

Curry, R.; Hollingworth, W.; Ellenbogen, R.G; Vavilala, M.S (2008). Incidence of hypo- and hypercarbia in severe traumatic brain injury before and after 2003 pediatric guidelines. *Pediatric Critical Care Medicine* (Lippincott Williams & Wilkins) 9 (2): 141–146.

Tasker, RC (2008). "Head and spinal cord trauma". In Nichols DG. Roger's Textbook of Pediatric Intensive Care (4th ed.). PA: Lippincott Williams & Wilkins. pp. 887–911. ISBN 978-0-7817-8275-3.

Marshall LF (2000). Head injury: Recent past, present, and future. *Neurosurgery* 47 (3): 546–61.

(eds.), John P. Adams, Dominic Bell, Justin McKinlay (2010). Neurocritical care : a guide to practical management. London: Springer. pp. 77–88. ISBN 978-1-84882-069-2.

O'Leary, R.; McKinlay, J. (2011). Neurogenic pulmonary oedema.

Continuing Education in Anaesthesia, Critical Care & Pain 11 (3): 87–92.

Strategies for Managing Multisystem Disorders. Hagerstwon, MD: Lippincott Williams & Wilkins. 2005. p. 370. ISBN 1-58255-423-4. http://books.google.com/?id=SPPiZWC8FpEC&pg=PA365&dq=%22traumatic+brain+injury%22+accident+sports.

Dunn IF, Ellegala DB (2008). "Decompressive hemicraniectomy in the management of severe traumatic brain injury". In Bhardwaj A, Ellegala DB, Kirsch JR. Acute Brain and Spinal Cord Injury: Evolving Paradigms and Management. Informa Health Care. pp. 1–2. ISBN 1-4200-4794-9.

Turner-Stokes L, Disler PB, Nair A, Wade DT (2005). Turner-Stokes, Lynne. ed. Multi-disciplinary rehabilitation for acquired brain injury in adults of working age. *Cochrane Database of Systematic Reviews* (3): CD004170.

McMillan TM, Oddy M (2000). "Service provision for social disability and handicap after acquired brain injury". In Wood RL, McMillan TM. Neurobehavioural Disability and Social Handicap Following Traumatic Brain Injury. East Sussex: Psychology Press. pp. 267–68. ISBN 0-86377-889-5

Deb S Crownshaw T (2004). The role of pharmacotherapy in the management of behaviour disorders in traumatic brain injury patients. *Brain Injury* 18 (1): 1–31.

Agrawal A, Timothy J, Pandit L, Manju M (July 2006). "Post-traumatic epilepsy: An overview". *Clin Neurol Neurosurg* 108 (5): 433–9.

Rao V, Lyketsos C (2000). Neuropsychiatric sequelae of traumatic brain Injury. *Psychosomatics* 41 (2): 95–103.

Brown AW, Elovic EP, Kothari S, Flanagan SR, Kwasnica C (March 2008). Congenital and acquired brain injury. 1. Epidemiology, pathophysiology, prognostication, innovative treatments, and prevention. *Archives of Physical Medicine and Rehabilitation* 89 (3 Supplement 1): S3–8.

Frey LC (2003). "Epidemiology of posttraumatic epilepsy: A critical review". Epilepsia 44 (Supplement 10): 11–17. doi:10.1046/j.1528-1157.44.s10.4.x.   PMID   14511389.   http://www.blackwell-synergy.com/doi/full/10.1046/j.1528-1157.44.s10.4.x?prevSearch=allfield%3A%28concussive%29.

Armin SS, Colohan AR, Zhang JH (June 2006). Traumatic subarachnoid hemorrhage: Our current understanding and its evolution over the past half century. *Neurological Research* 28 (4): 445–52. doi:10.1179/016164106X115053. PMID 16759448.

Agrawal A, Timothy J, Pandit L, Manju M (2006). "Post-traumatic epilepsy: An overview". Clinical Neurology and Neurosurgery 108 (5): 433–439.

"Persistent vegetative state" at Dorland's Medical Dictionary

Schiff ND, Plum F, Rezai AR (March 2002). "Developing prosthetics to treat cognitive disabilities resulting from acquired brain injuries". Neurological Research 24 (2): 116–24.

Kwasnica C, Brown AW, Elovic EP, Kothari S, Flanagan SR (March 2008). "Congenital and acquired brain injury. 3. Spectrum of the acquired brain injury population". Archives of Physical Medicine and Rehabilitation 89 (3 Suppl 1): S15–20.

Oliveros-Juste A, Bertol V, Oliveros-Cid A (2002). "Preventive prophylactic treatment in posttraumatic epilepsy" (in Spanish; Castilian). Revista de Neurolología 34 (5): 448–459.

Aimaretti, G.; Ghigo, E.(2007) Should Every Patient with Traumatic Brain Injury be referred to an Endocrinologist, Nature Clinical Practice of Endocrinology and Metabolism (2007)3:(4): 318–319 http://www.medscape.com/viewarticle/555088

Arlinghaus KA, Shoaib AM, Price TRP (2005). "Neuropsychiatric assessment". In Silver JM, McAllister TW, Yudofsky SC. Textbook of Traumatic Brain Injury. Washington, DC: American Psychiatric Association. pp. 59–62. ISBN 1-58562-105-6.

Hall RC, Hall RC, Chapman MJ (2005). "Definition, diagnosis, and forensic implications of postconcussional syndrome". Psychosomatics 46 (3): 195–202. http://psy.psychiatryonline.org/cgi/content/full/46/3/195.

Jordan BD (2000). "Chronic traumatic brain injury associated with boxing". Seminars in Neurology 20 (2): 179–85.

Mendez MF (1995). "The neuropsychiatric aspects of boxing". International Journal of Psychiatry in Medicine 25 (3): 249–62.

Draper, K; Ponsford, J (2009). "Long-term outcome following traumatic brain injury: A comparison of subjective reports by those injured and their relatives". Neuropsychological Rehabilitation 19 (5): 645–661. doi:10.1080/17405620802613935. PMID 19629849.

Rutherford, GW; Corrigan, JD (2009). "Long-term Consequences of Traumatic Brain Injury". Journal of Head Trauma Rehabilitation 24 (6): 421–423.

Temkin, N; Corrigan, JD, Dikmen, SS, Machamer, J (2009). "Social functioning after traumatic brain injury". Journal of Head Trauma Rehabilitation 24 (6): 460–467.

Velikonja, D; Warriner, E, Brum, C (2009). "Profiles of emotional and behavioral sequelae following acquired brain injury: Cluster analysis of the Personality Assessment Inventory". Journal of Clinical and Experimental Neuropsychology iFirst: 1–12.

Hawley, LA; Newman, JK (2010). "Group interactive structured treatment (GIST): a social competence intervention for individuals with brain injury". Brain Injury 24 (11): 1292–97. doi:10.3109/02699052.2010.506866.

Hesdorffer, DC; Rauch, SL, Tamminga, CA (2009). Long-term Psychiatric Outcomes Following Traumatic Brain Injury: A Review of the Literature. 24. pp. 452–459.

Fleminger S. Neuropsychiatric effects of traumatic brain injury. Psychiatr Times. 2010;27(3):40–45.

Dockree, PM, O'Keeffe, FM, Moloney, P, Bishara, AJ, Carton, S, Jacoby, LL, Robertson, IH (2006). "Capture by misleading information and its false acceptance in patients with traumatic brain injury". Brain 129 (1): 128–140.

Kreutzer, J, Stejskal, T, Ketchum, J, et. al. (March 2009). "A preliminary investigation of the brain injury family intervention: Impact on family members". Brain Injury 23 (6): 535–547.

Kreutzer, J, Kolokowsky-Hayner, S, Kemm, S & Meade, M (2002). "A structured approach to family intervention after brain injury". Journal of Head Trauma and Rehabilitation 17 (4): 349–367.

León-Carrión J, Domínguez-Morales Mdel R, Barroso y Martín JM, Murillo-Cabezas F (2005). "Epidemiology of traumatic brain injury and subarachnoid hemorrhage". Pituitary 8 (3–4): 197–202.

Alves OL, Bullock R (2001). "Excitotoxic damage in traumatic brain injury". In Clark RSB, Kochanek P. Brain injury. Boston: Kluwer Academic Publishers. p. 1. ISBN 0-7923-7532-7. http://books.google.com/?id=uRR4jKhF_iUC&pg=PA1&dq=Traumatic+brain+injury+causes. Retrieved 2008-11-28.

"Coma" at Dorland's Medical Dictionary

Cassidy JD, Carroll LJ, Peloso PM, Borg J, von Holst H, Holm L et al. (2004). "Incidence, risk factors and prevention of mild traumatic brain injury: Results of the WHO Collaborating Centre Task Force on Mild Traumatic Brain Injury". Journal of Rehabilitation Medicine 36 (Supplement 43): 28–60.

D'Ambrosio R, Perucca E (2004). "Epilepsy after head injury". Current Opinion in Neurology 17 (6): 731–735. doi:10.1097/00019052-200412000-00014. PMC 2672045. PMID 15542983. www.ncbi.nlm.nih.gov/pmc/articles/PMC2672045/.

Chesnutt RM, Eisenberg JM (1999). "Introduction and background". Rehabilitation for Traumatic Brain Injury. p. 9. ISBN 0-7881-8376-1.

Tolias C and Sgouros S (February 4, 2005). "Initial evaluation and management of CNS injury". eMedicine.com. http://www.emedicine.com/med/topic3216.htm. Retrieved 2007-12-16.

Carli P, Orliaguet G (February 2004). "Severe traumatic brain injury in children". Lancet 363 (9409): 584–85.

Necajauskaite, O; Endziniene M, Jureniene K (2005). "The prevalence, course and clinical features of post-concussion syndrome in children" (PDF). Medicina (Kaunas) 41 (6): 457–64. PMID 15998982. http://medicina.kmu.lt/0506/0506-01e.pdf.

Boake and Diller (2005). p.3

Granacher (2007). p.1

Scurlock JA, Andersen BR (2005). Diagnoses in Assyrian and Babylonian Medicine: Ancient Sources, Translations, and Modern Medical Analyses. Urbana: University of Illinois Press. p. 307. ISBN 0-252-02956-9. http://books.google.com/?id=alBmzfP3cpoC&pg=PA306&dq=%22head+trauma%22+ancient. Retrieved 2008-11-08.

Sanchez GM, Burridge AL (2007). "Decision making in head injury management in the Edwin Smith Papyrus". Neurosurgical Focus 23 (1): E5. http://thejns.org/doi/pdf/10.3171/FOC-07/07/E5.

Levin HS, Benton AL, Grossman R (1982). "Historical review of head injury". Neurobehavioral Consequences of Closed Head Injury. Oxford [Oxfordshire]: Oxford University Press. pp. 3–5. ISBN 0-19-503008-7. http://books.google.com/?id=EJJVT4ntacAC&pg=PA4&dq=%22head+trauma%22+ancient. Retrieved 2008-11-08.

Zillmer EA, Schneider J, Tinker J, Kaminaris CI (2006). "A history of sports-related concussions: A neuropsychological perspective". In Echemendia RJ. Sports Neuropsychology: Assessment and Management of Traumatic Brain Injury. New York: The Guilford Press. pp. 21–23. ISBN 1-57230-078-7. http://books.google.com/?id=dsEspzWGVzMC&pg=PA21&dq=%22traumatic+brain+injury%22+hippocrates. Retrieved 2008-10-31.

Corcoran C, McAlister TW, Malaspina D (2005). "Psychotic disorders". In Silver JM, McAllister TW, Yudofsky SC. Textbook Of Traumatic Brain Injury. Washington, DC: American Psychiatric Association. p. 213. ISBN 1-58562-105-6. http://books.google.com/?id=3CuM6MviwMAC&printsec=frontcover&dq=tarumatic+brain+injury. Retrieved 2008-11-08.

Eslinger, P. J.; Damasio, A. R. (1985). "Severe disturbance of higher cognition after bilateral frontal lobe ablation: patient EVR". Neurology 35 (12): 1731–1741.

Devinsky, O.; D'Esposito, M. (2004). Neurology of Cognitive and Behavioral Disorders (Vol. 68). New York: Oxford University Press. ISBN 978-0-19-513764-4 ISBN 0195137647.

Marion (1999). p.3.

Jones E, Fear NT, Wessely S (November 2007). "Shell shock and mild traumatic brain injury: A historical review". The American Journal of Psychiatry 164 (11): 1641–5. http://ajp.psychiatryonline.org/cgi/content/full/164/11/1641.

Belanger, H.G.; Kretzmer,T., Yoash-Gantz, R., Pickett, T, & Tupler, L.A (2009). Cognitive sequelae of blast-related versus other mechanisms of brain trauma. *Journal of the International Neuropsychological Society* 15 (1): 1–8.

Boake and Diller (2005). p.8

Bush, George H.W. (July 1990). "Project on the Decade of the Brain". http://www.loc.gov/loc/brain/proclaim.html. Retrieved September 30, 2009.

Yates D, Aktar R, Hill J; Guideline Development Group. (2007). "Assessment, investigation, and early management of head injury: Summary of NICE guidance". British Medical Journal 335 (7622): 719–20. doi:10.1136/bmj.39331.702951.47. PMC 2001047. PMID 17916856.
http://www.bmj.com/cgi/content/full/335/7622/719.

Bennett MH, Trytko B, Jonker B (2004). Bennett, Michael H. ed. Hyperbaric oxygen therapy for the adjunctive treatment of traumatic brain injury. *Cochrane Database of Systematic Reviews* (4): CD004609.

Barrett KF, Masel B, Patterson J, Scheibel RS, Corson KP, Mader JT (2004). "Regional CBF in chronic stable TBI treated with hyperbaric oxygen". Undersea & Hyperbaric Medicine: Journal of the Undersea and Hyperbaric Medical Society, Inc 31 (4): 395–406. PMID 15686271. http://archive.rubicon-foundation.org/4024. Retrieved 2008-10-30.

Freiberger JJ, Suliman HB, Sheng H, McAdoo J, Piantadosi CA, Warner DS (February 2006). "A comparison of hyperbaric oxygen versus hypoxic cerebral preconditioning in neonatal rats". Brain Research 1075 (1): 213–22.

Liu Z, Jiao QF, You C, Che YJ, Su FZ (June 2006). "Effect of hyperbaric oxygen on cytochrome C, Bcl-2 and Bax expression after experimental traumatic brain injury in rats". Chinese Journal of Traumatology (Zhonghua chuang shang za zhi), Chinese Medical Association 9 (3): 168–74.

Rockswold SB, Rockswold GL, Defillo A (2007). Hyperbaric oxygen in traumatic brain injury. *Neurological Research* 29 (2): 162–72.

Hardy P, Johnston KM, De Beaumont L, Montgomery DL, Lecomte JM, Soucy JP, Bourbonnais D, Lassonde M (2007). Pilot case study of the therapeutic potential of hyperbaric oxygen therapy on chronic brain injury. *Journal of the Neurological Science*s 253 (1–2): 94–105.

Appendix: Selected Online References Dealing With Different Forms of Traumatic Brain Injury (TBI)

Brief Notes

TBI is a major cause of death and disability worldwide, especially in children and young adults. Causes include falls, vehicle accidents, and violence. Prevention measures include use of technology to protect those suffering from automobile accidents, such as seatbelts and sports or motorcycle helmets, as well as efforts to reduce the number of automobile accidents, such as safety education programs and enforcement of traffic laws.

Brain trauma can be caused by a direct impact or by acceleration alone. In addition to the damage caused at the moment of injury, brain trauma causes secondary injury, a variety of events that take place in the minutes and days following the injury. These processes, which include alterations in cerebral blood flow and the pressure within the skull, contribute substantially to the damage from the initial injury.

TBI can cause a host of physical, cognitive, social, emotional, and behavioral effects, and outcome can range from complete recovery to permanent disability or death. The 20th century saw critical developments in diagnosis and treatment that decreased death rates and improved outcome. Some of the current imaging techniques used for diagnosis and treatment include CT scans computed tomography and MRIs magnetic resonance imaging. Depending on the injury, treatment required may be minimal or may include interventions such as medications, emergency surgery or surgery years later. Physical therapy, speech therapy, recreation therapy, and occupational therapy may be employed for rehabilitation.

Traumatic Brain Injury is defined as damage to the brain resulting from external mechanical force, such as rapid acceleration or deceleration, impact, blast waves, or penetration by a projectile. Brain function is temporarily or permanently impaired and structural damage may or may not be detectable with current technology.

TBI is one of two subsets of acquired brain injury (brain damage that occurs after birth); the other subset is non-traumatic brain injury, which does not involve external mechanical force (examples include stroke and infection). All traumatic brain injuries are head injuries, but the latter term may also refer to injury to other parts of the head However, the terms head injury and brain injury are often used interchangeably Similarly, brain injuries fall under the classification of central nervous system injuries and neurotrauma. In neuropsychology research literature, in general the term "traumatic brain injury" is used to refer to non penetrating traumatic brain injuries.

TBI is usually classified based on severity, anatomical features of the injury, and the mechanism (the causative forces). Mechanism-related classification divides TBI into closed and penetrating head injury. A closed (also called nonpenetrating, or blunt) injury occurs when the brain is not exposed. A penetrating (or open) head injury occurs when an object pierces the skull and breaches the dura mater, the outermost membrane surrounding the brain.

References:

Rehman T, Ali R, Tawil I, Yonas H (2008). "Rapid progression of traumatic bifrontal contusions to transtentorial herniation: A case report". Cases journal 1 (1): 203. http://www.casesjournal.com/content/1/1/203.

Jennett B (May 1998). "Epidemiology of head injury". Archives of Disease in Childhood 78 (5): 403–06. http://adc.bmj.com/cgi/content/full/78/5/403.

Saatman KE, Duhaime AC Workshop Scientific Team Advisory Panel Members et al. (2008). Classification of traumatic brain injury for targeted therapies. *Journal of Neurotrauma* 25 (7): 719–38 http:.//www.ncbi.nlm.nih.gov/pmc/articles/PMC2721779/ .

Department of Defense and Department of Veterans Affairs (2008). "Traumatic Brain Injury Task Force". http://www.cdc.gov/nchs/data/icd9/Sep08TBI.pdf.

Bayly, P.V.; Cohen, T. S., Leister, E. P., Ajo, D., Leuthardt, E. C., & Genin, G. M. (2005). "Deformation of the human brain induced by mild acceleration". Journal of Neurotrauma 22 (8): 845–856. http://www.ncbi.nlm.nih.gov/pmc/articles/PMC2377024/.

"NINDS Traumatic Brain Injury Information Page". National Institute of Neurological Disorders and Stroke. 2008-09-15. http://www.ninds.nih.gov/disorders/tbi/tbi.htm. Retrieved 2008-10-27.

Kushner D (1998). Mild traumatic brain injury: Toward understanding manifestations and treatment. *Archives of Internal Medicine* 158 (15): 1617–24. http://archinte.highwire.org/cgi/content/full/158/15/1617

"Signs and Symptoms". Centers for Disease Control and Prevention, National Center for Injury Prevention and Control. 2007-07-07. http://www.cdc.gov/ncipc/tbi/Signs_and_Symptoms.htm. Retrieved 2008-10-27.

Faul, M (2010). "Traumatic Brain Injury in the United States: Emergency Department Visits, Hospitalizations, and Deaths, 2002-2006".National Center for Injury Prevention and Control, Centers for Disease Control and Prevention. http://www.cdc.gov/traumaticbraininjury/tbi_ed.html#3. Retrieved August 9, 2011.

"Traumatic brain injury". Centers for Disease Control and Prevention, National Center for Injury Prevention and Control. 2007. http://www.cdc.gov/ncipc/factsheets/tbi.htm. Retrieved 2008-10-28.

Park E, Bell JD, Baker AJ (April 2008). Traumatic brain injury: Can the consequences be stopped?. *Canadian Medical Association Journal* 178 (9): 1163–70. //www.ncbi.nlm.nih.gov/pmc/articles/PMC2292762/

"Pentagon Told Congress It's Studying Brain-Damage Therapy".ProPublica. http://www.propublica.org/article/pentagon-told-congress-its-studying-brain-damage-therapy. Retrieved 2011-01-23. "Brave Americans who risked everything for their country and sustained traumatic brain injuries -- the signature injury of the wars in Iraq and Afghanistan -- deserve cognitive rehabilitation therapy to help them secure the best futures possible. It is unacceptable that the United States has been at war for nearly a decade and there is still no plan to treat these soldiers."

American Academy of Pediatrics: Committee on Child Abuse and Neglect (July 2001). Shaken baby syndrome: Rotational cranial injuries. Technical report. *Pediatrics* 108 (1): 206–10. http://pediatrics.aappublications.org/cgi/content/full/108/1/206

Morrison AL, King TM, Korell MA, Smialek JE, Troncoso JC (1998). Acceleration-deceleration injuries to the brain in blunt force trauma. *American Journal of Forensic Medical Pathology* 19 (2): 109–112. http://meta.wkhealth.com/pt/pt-core/template-journal/lwwgateway/media/landingpage.htm?issn=0195-7910&volume=19&issue=2&spage=109.

Poirier MP (2003). Concussions: Assessment, management, and recommendations for return to activity (abstract). *Clinical Pediatric Emergency Medicine* 4 (3): 179–185. http://www.sciencedirect.com/science?_ob=ArticleURL&_udi=B75BD-49H1C2F-7&_user=3356446&_origUdi=B6VDJ-44KHFBN-8&_fmt=high&_coverDate=09%2F30%2F2003&_rdoc=1&_orig=article&_acct=C000060332&_version=1&_urlVersion=0&_userid=3356446&md5=9c2a61c0c62684c26cf317a8ea637458.

Narayan RK, Michel ME, Ansell B et al. (May 2002). Clinical trials in head injury. *Journal of Neurotrauma* 19 (5): 503–57. www.ncbi.nlm.nih.gov/pmc/articles/PMC1462953/.

Coles JP (July 2007). Imaging after brain injury. *British Journal of Anaesthesia* 99 (1): 49–60. http://bja.oxfordjournals.org/cgi/pmidlookup?view=long&pmid=17573394

McCrory PR (August 2003). "Brain injury and heading in soccer". BMJ 327   (7411):   351–52. http://www.bmj.com/cgi/content/full/327/7411/351?etoc.

McIntosh AS, McCrory P (June 2005). Preventing head and neck injury. *British Journal of Sports Medicine* (Free registration required) 39 (6): 314–18. http://bjsm.bmj.com/cgi/content/full/39/6/314.

Kluger, Jeffrey. "Dealing with Brain Injuries. Time Magazine, April 6, 2009, p. 57. Online: http://www.time.com/time/magazine/article/0,9171,1887856,00.html. Accessed: May 1, 2009

Office of Communications and Public Liaison (February 2002).

"Traumatic brain injury: Hope through research". NIH Publication No. 02-2478. National Institute of Neurological Disorders and Stroke, National Institutes of Health. http://www.ninds.nih.gov/disorders/tbi/detail_tbi.htm. Retrieved 2008-08-17. "Many patients with mild to moderate head injuries who experience cognitive deficits become easily confused or distracted and have problems with concentration and attention. They also have problems with higher level, so-called executive functions, such as planning, organizing, abstract reasoning, problem solving, and making judgments, which may make it difficult to resume pre-injury work-related activities. Recovery from cognitive deficits is greatest within the first 6 months after the injury and more gradual after that."

Moppett IK (July 2007). Traumatic brain injury: Assessment, resuscitation and early managemen". *British Journal of Anaesthesiology* 99 (1): 18–31.http://bja.oxfordjournals.org/cgi/content/full/99/1/18.

Roberts I, Smith R, Evans S (February 2007). Doubts over head injury studies. 334 (7590): 392–94. http://www.bmj.com/cgi/content/full/334/7590/392.

Frey LC (2003). "Epidemiology of posttraumatic epilepsy: critical review". Epilepsia 44 (Supplement 10): 11–17. http://www.blackwell-synergy.com/doi/full/10.1046/j.1528-1157.44.s10.4.x?prevSearch=allfield%3A%28concussive%29.

Aimaretti, G.; Ghigo, E.(2007) Should every patient with traumatic brain injury be referred to an endocrinologist, *Nature Clinical Practice of Endocrinology and Metabolism* (2007)3:(4): 318–319 http://www.medscape.com/viewarticle/555088

Hall RC, Hall RC, Chapman MJ (2005). "Definition, diagnosis, and forensic implications of postconcussional syndrome". Psychosomatics 46 (3): 195–202. doi:10.1176/appi.psy.46.3.195. PMID 15883140.http://psy.psychiatryonline.org/cgi/content/full/46/3/195

About the Author

George Doherty resides in Laramie, WY where he founded the Rocky Mountain Region Disaster Mental Health Institute, Inc. He is currently employed as the President/CEO of this organization and also serves as Clinical Coordinator of the Snowy Range Critical Incident Stress Management Team.

He has been involved with disaster relief since 1995, serving as a Disaster Mental Health Specialist with such incidents as the UP train wreck in Laramie, Hurricane Fran in North Carolina, the Cincinnati floods in Falmouth, KY and Tropical Storm Allison in Southeast Texas. He served as Supervisor for Disaster Mental Health for flash floods in Ft. Collins, and spent a month as the Red Cross Disaster Mental Health Coordinator for western Puerto Rico in the aftermath of Hurricane George.

He has also published numerous articles in disaster mental health and traumatic stress publications and served as Guest Editor for two Special Editions of the journal *Traumatology* (1999 & 2004).

He served as an officer in the US Air Force and was an OTS instructor, squadron commander and other positions. Additionally, he served 11 years involved in Air Search & Rescue with Civil Air Patrol (US Air Force Auxiliary) in WY as Squadron Commander, Deputy Wing Commander, Air Operations Officer, and Master Observer. He is a Certified Instructor with the Wyoming Peace Officers Standards and Training (POST).

He has extensive experience conducting CISM debriefings with first responders and others and is a member of a national crisis care network, providing assistance to companies and other organizations following critical incidents involving sudden deaths and similar traumatic events.

He is a Licensed Professional Counselor in private practice and has been an adjunct instructor for a number of colleges, including Northern Nevada Community College and the University of Wyoming.

Organizational memberships include the American Counseling Association, Voting Associate Member of the American Psychological Association, American Academy of Experts in Traumatic Stress (AETS), Association of Traumatic Stress Specialists (ATSS), Traumatic Incident Reduction Association (TIRA), Certificate of Specialized Training in the field of Mass Disaster and Terrorism, Wyoming Department of Health Hospital Emergency Preparedness Advisory Committee; Research Advisor and Research Fellow: American Biographical Institute (ABI), Editorial Advisory Board Member and Book Reviewer: PsyCritiques (*APA Journal*), Life Member of the Air Force Association, Life Member of the Military Officers Association of America, Member American Legion, Life Member: Pennsylvania State University Alumni Association, Alumni Admissions Volunteer - Pennsylvania State University.

Publications include: Crisis *Intervention Training for Disaster Workers: An Introduction.*; Editor and contributor for the *Proceedings of Rocky Mountain Region Disaster Mental Health Conferences* (2005, 2006, 2007, 2008). Served as Guest Editor for Special issues of the journal *Traumatology* on Disaster Mental Health (1999) and Crises in Rural America (2004); Cross-cultural Counseling in Disaster Settings. - *Australasian Journal of Disaster and Trauma Studies* (1999). Published reviews include: Understanding Oslo in Troubled Times; Responders to September 11, 2001: Counseling: Innovative Responses to 9/11 Firefighters, Families, and Communities; Genocide: A Human Condition?; Stress Management, Wellness and Organizational Health; Leadership Competency and Conflict; Leadership: Lessons from the Ancient World - all in PsyCritiques. Conference Director for annual Rocky Mountain Disaster Mental Health Conferences 1999 -present.

Additional past positions include: Masters Level Psychologist – Rural Clinics (State of Nevada), 1980 – 1986; Veterans Counselor (VA Contract) – 1980-1986, NV; Counselor – pre-delinquent children and families – CORA Services (Philadelphia, PA).1972-1975; Program Coordinator – Community Action Programs (EOAC, Office of Economic Opportunity – Waco/McLennan County, TX) 1968-1971.

Current Courses/Workshops he teaches include:

- Crisis Intervention In Disasters: Training For Workers - An Introduction - 12 CEU
- CISM: Individual Crisis Intervention And Peer Support

- CISM: Group Crisis Intervention
- CISM: Advanced Group Crisis Intervention
- CISM: Strategic Response to Crisis (Capstone Course)
- Return To Equilibrium: Disaster Mental Health- 4 CEU
- Return To Equilibrium: Returning Military And Families - 8 CEU
- From Crisis to Recovery: Strategic Planning for Response, Resilience and Recovery - 12 CEU

Email: rmrdmhi@gmail.com

.

Index

Who will step up to meet the challenge of the next rural crisis?

Rural practice presents important yet challenging issues for psychology, especially given uneven population distribution, high levels of need, limited availability of rural services, and ongoing migration to urban centers. It is critical that mental health professionals and first responders in rural areas become aware of recent research, training and approaches to crisis intervention, traumatology, compassion fatigue, disaster mental health, critical incident stress management, post-traumatic stress and related areas in rural environments. Critical issues facing rural areas include:

George W. Doherty, MS, LPC

Crisis In The American Heartland –
Disasters & Mental Health
In Rural Environments:
An Introduction
(Volume 1)

- Physical issues such as land, air, and water resources, cheap food policy, chemicals and pesticides, animal rights, corruption in food marketing and distribution, and land appropriation for energy development.
- Quality of life issues such as rural America's declining share of national wealth, problems of hunger, education, and rural poverty among rural populations of farmers and ranchers.
- Direct service issues include the need to accommodate a wide variety of mental health difficulties, client privacy and boundaries, and practical challenges.
- Indirect service issues include the greater need for diverse professional activities, collaborative work with professionals having different orientations and beliefs, program development and evaluation, and conducting research with few mentors or peer collaborators.
- Professional training and development issues include lack of specialized relevant courses and placements.
- Personal issues include limited opportunities for recreation, culture, and lack of privacy.

Doherty's first volume in this new series *Crisis in the American Heartland* explores these and many other issues. Each volume available in trade paper, hardcover, and eBook formats.

ISBN 978-1-61599-075-7 (trade paper)
For more information please visit www.RMRInstitute.org

www.ingramcontent.com/pod-product-compliance
Lightning Source LLC
Chambersburg PA
CBHW080758300326
41914CB00055B/939